# *Improving Utilities with Systems Thinking: People, [illegible] chnology*

Zdenko Vitasovic, Gustaf Olsson, Pernille Ingildsen, & Scott Haskins

2022

Water Environment Federation

601 Wythe Street

Alexandria, VA 22314–1994 USA

https://www.wef.org

International Water Association

Republic–Export Building, Units 1.04 & 1.05

1 Clove Crescent

London, E14 2BA, UK

https://www.iwapublishing.com

ISBN 978-1-57278-429-1 WEF
ISBN 9781789063134 IWAP

**IMPORTANT NOTICE**

The material presented in this publication has been prepared in accordance with generally recognized engineering principles and practices and is for general information only. This information should not be used without first securing competent advice from an industry professional with respect to its suitability for any general or specific application.

No reference made in this publication to any specific method, product, process, or service constitutes or implies an endorsement, recommendation, or warranty thereof by WEF or IWA.

WEF and IWA make no representation or warranty of any kind, whether expressed or implied, concerning the accuracy, product, or process discussed in this publication and assumes no liability.

Anyone using this information assumes all liability arising from such use, including but not limited to infringement of any patent or patents.

WEF and IWA work hard to ensure that the information in this publication is accurate and complete. However, it is possible that the information may change after publication, and/or that errors or omissions may occur. We welcome your suggestions on how to improve this publication and correct errors. WEF and IWA disclaim all liability for, damages of any kind arising out of use, reference to, or reliance on information included in this publication to the full extent provided by state and Federal law.

## About WEF

The Water Environment Federation (WEF) is a not-for-profit technical and educational organization of 30,000 individual members and 75 affiliated Member Associations representing water quality professionals around the world. Since 1928, WEF and its members have protected public health and the environment. As a global water sector leader, our mission is to connect water professionals; enrich the expertise of water professionals; increase the awareness of the impact and value of water; and provide a platform for water sector innovation. To learn more, visit www.wef.org.

## About IWAP

The International Water Association (IWA) is a network and an international global knowledge hub open to all water professionals and anyone committed to the future of water. With its legacy of over seventy years, it connects water professionals around the world to find solutions to global water challenges as part of a broader sustainability agenda.

As a non-profit organization with members in more than 130 countries, IWA connects scientists with professionals and communities so that pioneering research offers sustainable solutions. In addition, the association promotes and supports technological innovation and best practices through international frameworks and standards. Each year, IWA organizes and sponsors over 40 specialised conferences and seminars on a wide variety of water and sanitation topics worldwide. Further, IWA publishes 14 scientific journals and 30+ books per year by IWA Publishing. Meet us at the IWA World Water Congress & Exhibition, the IWA Water and Development Congress & Exhibition, and our Specialist Groups Conferences worldwide. For more information, please visit www.iwa-network.org.

## Introduction of Authors

| | |
|---|---|
| Zdenko (Cello) Vitasovic, PhD, PE<br><br>9D Analytics, (corresponding author), cello@9danalytics.com<br><br>Principal Investigator, WRF 4806 Project (UAIM)<br><br>Co-Principal Investigator, WRF Project 5039 (Smart Utilities and Intelligent Water Systems)<br><br>Past Chair of WEF Intelligent Water Systems Committee<br><br>Recipient of the WEF Harrison Prescott Eddy Medal for outstanding Research (1998)<br><br>His focus has been on mathematical modeling, operational technology and real-time controls, and information systems. He has executed a number of projects for water sector utilities in North America, UK, China, Hong Kong, Singapore, and Australia. |  |
| Gustaf Olsson<br><br>Lund University, Lund, Sweden, gustafolsson3@gmail.com<br><br>Professor in Industrial Automation and since 2006 professor emeritus at Lund University, Sweden. He has dedicated his research to control and automation of urban water systems, power production, electrical power systems, and industrial processes. During the last decade he has devoted most of his research on the water-energy nexus.<br><br>In 2010 he received the International Water Association (IWA) Publication Award. In 2012 he was the awardee of an Honorary Membership of IWA as well as Honorary Doctor degree at the Technical University of Malaysia. In 2014 he was appointed Distinguished Fellow of the IWA. |  |

<table>
<tr><td>He has been part-time guest professor during more than a decade at Tsinghua University, Beijing, and at the Technical University of Malaysia. He is honorary faculty member of Exeter University, UK, and advisor to several international research groups and programs.<br><br>Gustaf has authored 12 international books—some of them translated into Russian, German, Korean, and Chinese—and contributed with chapters in another 21 books as well as about 200 scientific publications. His most recent book, Clean Water Using Solar and Wind—Outside the Power Grid (2018) is now available open access.</td><td></td></tr>
<tr><td>Pernille Ingildsen<br><br>Hillerød Forsyning, Hillerød, Denmark, pi@hfors.dk<br><br>Aa water utility practitioner with one foot in the academic world. She has authored Get the Most Out of Your Wastewater Treatment Plant, Smart Water Utilities (open access), and Water Stewardship (open access)—water is truly a heart's matter.<br><br>Her main contribution is combining theoretical concepts within sustainability and intelligent control and implementing them in practice—and in that way making sure that both the theoretical and the practical domain learns from the other.<br><br>As a water professional, she has been working in water utilities, in consulting companies, in product companies, in start-ups, and in universities.</td><td></td></tr>
</table>

Scott Haskins

Co-Principal Investigator, WRF 4806 Project (UAIM)

Mr. Haskins is a subject matter expert and industry leader in utility and financial management, asset and risk management, contracting, social equity, operations and maintenance, strategic planning, leadership and workforce development, performance management and benchmarking, environmental management, resilience, triple bottom line, and business case analysis. His background includes 30+ years of utility and municipal leadership experience, primarily with City of Seattle, directing finance, project development, operations and maintenance, community programs, planning and executive leadership over the full range of wastewater, drinking water, and solid waste functions. He has also performed leadership and consulting services for CH2M and Jacobs Engineering, including for benchmarking and a wide range of strategic, leadership, and operations services.

Mr. Haskins actively participates in the water and wastewater utility industry, serving in many leadership roles; has participated in numerous research projects for national and international organizations; has worked extensively on domestic and capacity building assignments; has authored many journal articles and conference presentations; and is co-author of several books. He also serves on advisory boards for Seattle City Light, and the Evans School of Public Policy and Governance at University of Washington, and was a long-time member of EPA's Environmental Finance Advisory Board.

Photo courtesy of Karen Orders Photography.

## Contents

# Foreword

Barry Liner, PhD, PE, BCEE

*Chief Technical Officer, The Water Environment Federation, Alexandria, Virginia, USA*

At their core, water utilities have a noble mission for literally centuries: protect public and environmental health through stewardship of all water resources including drinking water, wastewater, and stormwater. Because of their critical mission, most water utilities are in the public sector and operate as monopolies in their service areas. While private sector businesses must focus on their competitive advantage, water utilities' (whether public or private) competitive advantage is the public good. In the 21st century, utilities are facing challenges like funding for ageing infrastructure, workforce transformation, economic development, and technological innovation. Utility management professionals can use some techniques from business to improve operations, enhance efficiency, and instill a culture of embracing innovation.

The Water Intrapreneurs for Successful Enterprises (WISE) movement, which evolved from the Utility Analysis and Improvement Methodology (UAIM) initiative, provides a way of thinking about the business of running a water utility. Because value creation at a water utility has nuances that only other entities serving the public good would fully appreciate, the WISE framework builds from expertise from utility managers worldwide. This publication explains the benefits of system thinking and maturity models, and how people, process, and technology interact. The business process mapping and other tools of the WISE framework enable utility management professionals at all levels to apply the best of business process improvement practices in a way that is tailored to the unique demands of value creation at water utilities. Watching the evolution of the movement from a handful of people developing UAIM to the tried and tested tools in the WISE framework has been fascinating. I look forward to the continued evolution and impact of the WISE community.

# Prologue: From Smart to WISE—A Wake-Up Call

People in the water sector have the honor and privilege to work on a noble mission as stewards of a fundamental component of life: water. Water sector utilities are facing several challenges; each of us needs to consider if we are fulfilling all aspects of our responsibilities as water professionals, servants to our communities, and human beings that are part of nature.

Our sector is facing different categories of challenges:

1. Category 1—challenges in our role as *purveyors of water and wastewater services*. The focus of this challenge is on performing within a role that utilities have traditionally perceived as their "core mission" that includes efficient execution of tasks such as securing adequate quantities of water from the environment, providing water to their customers, collecting and treating wastewater, and managing stormwater. To meet these challenges, we need to implement methodologies that allow us to make decisions based on sound engineering and scientific knowledge.
2. Category 2—challenges in our role as *stewards and protectors of a precious resource* that has a profound impact on the sustainability of the cycle of life as we know it. To meet this category of challenges, we must ensure that our utility enables the local human community to become a seamless and integral part of nature. For a sustainable future, we need to minimize and avoid detrimental environmental effects, long term or short term.
3. Category 3—challenges in our role as *leaders, managers, and co-workers* in our organizations. We are responsible for creating and supporting organizational culture that is based on sound values and allows people working in water sector utilities to fully engage in the overall organizational mission and achieve their full potential.

In each of these categories, our ability to meet these challenges is controlled and constrained by our underlying assumptions and attitudes and by our mindset and mental models that define our approach to problems and solutions. Our thoughts and actions have been shaped by dominant paradigms that include the following:

a) The vast majority of water sector utilities are in the public sector, and their organizations are governed and managed based on the paradigm that originates from the traditional military (Laloux, 2014). The *"military" paradigm* includes hierarchical structure, top-down command and control decision making, and static/fixed view of roles and responsibilities.

b) In relatively recent history, humans have been perceiving nature as something that needs to be conquered and used to satisfy the needs of humans. Environment has been perceived as an inexhaustible source of natural wealth that should be extracted for short-term gain, and humans have been seen as a dominating force separate from nature. The world has been perceived within a *paradigm of a machine* that serves the needs of humans.
c) For the past four hundred years, the dominant paradigm for our learning has been *structural decomposition*: to break down the complex mosaic of the phenomena that we are trying to understand into *fragments* and use highly specialized knowledge to study the individual tiles.
d) Even when looking at themselves, humans have created a *division between the body and the mind*, treating them as separate and independent components.

"Ah, not to be cut off,
not through the slightest partition
shut out from the law of the stars.
The inner—what is it?
if not the intensified sky,
hurled through with birds and deep
with the winds of homecoming."

Rainer Maria Rilke, Ahead of All Parting: The Selected Poetry and Prose, Modern Library, 1995

Changing the dominant paradigm that views the world as a machine is difficult because we have all been raised in this mindset. The high-level goal of Water Intrapreneurs for Successful Enterprises (WISE) is to adopt, apply, and maintain a *systemic view* that is profoundly different from the "machine" paradigm that has been dominant for the past several generations.

There is abundant evidence that these dominant paradigms are having an adverse effect on our ability to address all categories of our challenges. The degradation of the environment has taken us to the boundaries of what Earth can take—or what our watersheds can take. We will not be able to fix the damage to the environment using the same paradigm. As we diligently extract non-renewable resources, we are also greatly diminishing and endangering the very system that created life on this planet.

The machine thinking has mindlessly destroyed so much; we need a different mindset to repair the damage. Water sector utilities, as stewards of a precious natural resource, have an important role to play in the healing and restoration of our world. We can heal the damage that is being done by these omnipresent but outdated mindsets. It is time to embrace a mindset based on *systems thinking*, where the diverse networks, connections, and interactions rather than parts in isolation become our main focus. Systems thinking and the living network paradigm enable us to see our organizations, ourselves, our communities, and our environment as components of a larger ecosystem and to recognize the importance of interactions among different systems.

You are not a box on an organization chart. You are more than a set of organs within your body. A bee or a flower is not a machine, nor are rivers, forests, air, or oceans. Life is precious and more complex, and we need a mindset and practical methods that recognize this complexity.

Change is achievable, and the WISE "movement" includes a vision for such transformation for the water sector. We propose some initial and modest practical steps that we can take to improve organizational and personal learning. WISE is a highly collaborative effort that is driven by active engagement of staff and managers from different water sector utilities. The wisdom of WISE is in the employees from participating utilities, in their imaginative power, in their knowledge, in their passion, and in their commitment. Ernst Schumacher coined the concept of "economy as if people mattered"; WISE is a journey to improve utilities using a holistic approach that recognizes that many aspects matter, including people, the work that they do, and their responsibilities to help each other and protect nature.

Utilities participating in the WEF WISE program will continue the work that was started by WRF's UAIM (Utility Analysis and Improvement Methodology) project to collaboratively develop and apply a methodology based on systems thinking to describe, analyze, and improve the performance of water sector utilities. This is a holistic, value-based approach focused on real challenges faced by utilities. The practicality of this approach is being documented in a number of case studies.

Changing a dominant paradigm is difficult; Capra and Luisi wrote in their book "Systems View of Life" that a change of perspective from machine-based to systems and interactions-based thinking is not driven by logic alone but has a strong emotional–relational sphere. We want to do well, but we also desire to do good; when we stop viewing the world as a machine, systems thinking allows us to see the world in a way that is much more in tune with the complex nature of our challenges.
The spirit of WISE is reflected in a few common core values and beliefs that inspire our mission:

1. As professionals in the water sector, we recognize that we are responsible for meeting both the short-term and long-term needs of our stakeholders, customers, and local communities.

2. As scientists and engineers, we are aware of the threat to all life and the degradation of the environment caused by human activities.
3. As employees and managers, we are keen to enable continuous learning and apply a holistic approach to address challenges in our organizations.
4. As colleagues and co-workers, we are keen to enable effective collaboration among different parts of our organization.
5. As members of a broader community of professionals engaged in different water sector utilities, we are keen to engage in peer-to-peer collaboration among utilities.
6. As human beings, we want to help create and sustain an organizational culture in our utilities that is based on mutual trust and respect.

Many organizations start with a good purpose but, over time, the conservation of the organization becomes more important than the original purpose—or internal competition for power positions becomes peoples' focus. Our primary mission is to highlight and focus on the original foundational core goals of water sector utilities to protect a wondrous, life-sustaining substance. We are often challenged by the daily struggles with frustrating bureaucracy, difficult people, changing regulations, and so on. Focusing on our mission, remembering our core values and beliefs, working together with our colleagues and peers, and sharing a common purpose will help us avoid the loss of joy and pride in our work.

Probably something deep inside you feels skeptical—perhaps excited but cautious; this is understandable because the challenges appear to be enormous. The change seems very difficult or impossible: the "machine way of viewing life" has been dominant for a few hundred years, it has been accepted as "natural," and it is difficult to imagine that this could change. However, looking at human life from an evolutionary perspective of millions of years, it becomes clear that the machine view is very new, experimental, and not "natural." The "machine" view of the world has existed only for a few hundred years, and it has led us to the current crisis. We cannot solve the current crisis with the same paradigm that created it. We cannot go back in time, but we can choose to go forward wisely. Our choices will determine if we make the world better for future generations.

We will need the courage to do "the right thing" and creativity to find new ways and methods to achieve the needed transformation. We will need to look at nature not as something that needs to be conquered and exploited; it should be seen as a partner in life. If we remember that we are a part of nature it will become more clear that a battle with nature is a war against humanity.

The WISE effort focuses on developing frameworks, models, methods, and best practices that enable the transformation to improved water sector utilities. This is applied research that must produce actionable and practical deliverables. WISE will leverage system thinking to help organizations heal the damage caused by the dominant paradigms of machine and the military.

Ernst F. Schumacher puts it like this in "Small Is Beautiful": "Modern man does not experience himself as a part of nature but as an outside force destined to dominate and conquer it. He even talks of a battle with nature, forgetting that, if he won the battle, he would find himself on the losing side." We must attend to the living now and focus on what creates more life—in ourselves, among our fellow human beings, and among all the living creatures on Earth.

Water and wastewater utilities depend on all of the actors in the network, including law makers, municipalities, industries, nongovernmental organizations, consultants, contractors, researchers, students, manufacturers, and the public community. We cannot solve everything at once. We need a method and practical steps, but we also need a fundamental driving force that will push us through these steps. This driving force is most powerful if it comes deep down from within ourselves.

Carl Jung wrote "Who looks outside, dreams; who looks inside awakes." The WISE program considers the immediate challenges by providing a pragmatic and actionable structure to enable problem solving using a different paradigm. However, the force to drive these changes, to motivate people and companies, needs to come from within us. This is why our highest hope for this document is to serve as a wake-up call, to inspire the will for improvement and change, and to show that there is a practical way to heal our organizations and the environment. We can change our departments. We can change our work practices. We can change our technologies. We can change how we treat each other. Each small change at any given time will send ripples through the system and contribute to the overall transformation. The best way forward is through collaboration, allowing many to contribute a voice and add value. Each participant in WISE will bring their own instrument and ideas and sound their own notes. Together, all individual contributions will produce a regenerative, rich, sustainable music about the world and life. Come and join the band!

# Guide to the Reader

## Introducing the Document

This document builds on a collaborative endeavor during four years by more than twenty water sector utilities on a project that was initially started with Water Environment and Reuse Foundation (WE&RF) and continued in 2019 with the Water Research Foundation (WRF). The WRF project called Utility Analysis and Improvement Methodology (UAIM) will be completed in 2021. In early 2021, the current UAIM partners are as follows:

Baltimore City Department of Public Works, Pennsylvania

Charlotte Water, North Carolina

City of Grand Rapids, Michigan

Clean Water Services, Oregon

DC Water, Washington DC

Environment Agency, Team 2100, UK

Great Lakes Water Authority, Michigan

Gwinnett County, Georgia

KC Water, Kansas City, Missouri

King County Wastewater Division, Washington

Loudoun Water, Virginia

Louisville Metropolitan Sewer District, Kentucky

Metro Vancouver, British Columbia, Canada

Metropolitan Council Environmental Services, Twin Cities, Minnesota

Orange County, Florida

Portland Water, Oregon (water)

Portland Bureau of Environmental Services, Oregon (wastewater)

San Francisco Public Utilities Commission, California

Tacoma Water, Washington

Toho Water, Florida

VCS (VandCenter Syd), Denmark

Washington Suburban Sanitary Commission, Maryland

The effort will continue as a program called WISE (Water Intrapreneurs for Successful Enterprises) under the sponsorship of the Water Environment Federation (WEF).

UAIM has been a network of peer utilities collaborating to create industry-leading business practices and processes. The intention for this publication is to inform about the past and inspire about the future of this collaborative effort that includes a high level of active engagement by participating utilities; authors also hope that new utilities may be encouraged and inspired by the content and principles of this document to join this partnership. The WISE mission is to enable utility management practices focusing on performance and value to the organization, community, and environment, through application of systems thinking and a methodology for improvement.

Some short overviews and presentations of the WISE concept and approach have been presented earlier.[1]

## The Topic

We have examined management of water sector utilities from a "systems engineering" or a holistic point of view: starting with the complete system, then defining the components, considering the interactions between them, measuring key variables of value creation and impact, making adjustments, and learning. The overall goal is to establish a knowledge-based road map for improvement including practical methodologies based on science and grounded in the real needs of water sector utilities.

A key message in a WISE systems approach is that an integrated system has qualities that the sum of the individual components cannot explain. It is by a holistic view that the system—utility, community, and environment—can be learning and developing in a more sustainable way so that a much broader value can be created. It should appear not only in monetary profit but also as a better performance within the organization. This could be expressed in terms of an increased well-being for all participants or components of the system, including people, components, information systems, environment, customers, and infrastructure. Such well-being requires that the overall system is well designed and well balanced. In a system that is not well balanced, it will often be one or a few of the components that bears the brunt of the imbalance. In organizations with some stressed-out employees, it may be an environmental responsibility that is neglected or it may be customers that are not well served.

The outcome of the research has led us to formulate the *vision of a wise utility.* To appreciate how business processes, people, governance, and technology can ***interact*** to form a better water system is a demanding task and requires deep reflection to understand. It is the norm of *reciprocity* that holds most communities, societies, organizations, and companies together through time. Predictable relationships are much more powerful than "imposed" management orders. Reciprocity is the glue that governs the myriads of social exchanges that take place every day in an organization, most of which have nothing to do with the realm of formal management. It is vital that leaders build transparent and open communications and incentives to reward fair, equal, reciprocal, and resilient exchanges between people.

We hope that this publication will serve as an expanded conceptual framework for future continued efforts in water utility management.

| | |
|---|---|
| **The Intended Reader** | We hope to raise the interest among different groups of readers:<br><br>• *Decision makers* on different levels of utilities, who wish to get a holistic view of the truly complex interactions in the organization. One should keep in mind that every employee is making decisions every day. Any one person in an organization has some level of ability and responsibility to change the balance of the system.<br>• *Engineers* or *operators* who want to implement advanced technologies but struggle with "buy-in," acceptance, and adoption within the organization. How to communicate to different categories that the application of new technologies can solve key issues and be a central enabler for value creation?<br>• *Economists and financial professionals* who want to ensure optimal value creation for the means available, that is, optimal performance and value creation.<br>• *Human resources staff and management* who wish to find effective ways to affect the culture in the organization to enable the application of new technologies through influencing individuals' attitudes. |

| Key Messages | |
|---|---|
| | • Managing a utility is a complex problem that benefits from a methodology based on systems thinking. The silo mindset in management has to be abandoned. We need to better understand how groups can work together to ensure better performance, coherence, and well-being of the system. These groups should be able to adapt to the varying needs. Important interactions between people, technology, as well as value creation for utility, community, and nature must be understood and considered in the management.<br>• The WISE approach is based on trust, honesty, and transparency and considers how people execute business processes. Technology is one of the tools but only part of the solution. Cooperation between people, organizations, and technology is the key to create value.<br>• WISE methodology starts with assessments based on maturity models and includes specific steps for improvement that include change management and learning.<br>• WISE includes active engagement, collaboration, and sharing by the participating utilities.<br>• WISE is applied research; the participating utilities base the research priorities on their needs. |

# Executive Summary

This document starts by summarizing the findings of a research project to develop and apply a standard methodology for improving performance of water sector utilities. It then shifts focus to a vision for a new WISE project that will continue the collaborative efforts for improving management of utilities under the sponsorship of WEF. Most people think that smart technology is the solution, but technology alone is not enough to take utilities to the next "wise" level. Utilities also need to emphasize that people and processes should be subject to continuous learning. How does a wise utility proactively navigate external impacts on customers and the natural environment that our livability depends on? The symbiosis of people (organization and community), processes, technology, and environment is the ultimate recipe for value generation.

The value chain of the water sector is complex and includes a range of different aspects of the urban water cycle, from the water source to the receiving water. It is vital to incorporate the environment as part of this cycle. All of us are part of nature. This means that water extraction as well as the use of water effluent is not something outside our living domain; in the end, all of us will live with the consequences of our operations. In other words, the value chain includes both the internal value within the utility and the interaction with external entities and stakeholders including water customers, resources, and nature.

Utilities perform a range of different functions in widely different timescales ranging from operational to tactical and further to strategic operations and decisions. Automatic control systems act on equipment within minutes and hours, and process change decisions are performed in days and weeks. Establishing maintenance and asset management systems, as well as those for planning and execution of capital improvement projects, usually take years. The nine entries in the following 3×3 matrix (from Figure 1.2) illustrate a holistic view of internal value creation within a utility. A wise utility will consider each of the entries on their own as well as the interactions between them.

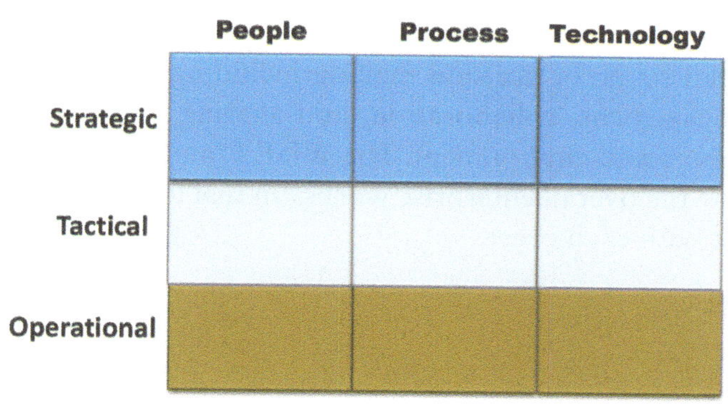

In Chapter 2, we aim to provide an overall vision, given current and future water sector challenges. The vision is an aspirational concept of a wise utility. The fact that water is life ought to influence all our attitudes and use of water. In our urbanized lives, we easily forget our connection to, and dependence on, nature. The balance between urban water use and nature is threatened by increasing population and urbanization, higher demand from industries and food production, intensified by climate change and more erratic precipitation patterns. The visions expressed in the document do not provide all the answers but should hopefully serve as guides in a continuous process improvement. The ambition is to encourage an ongoing dialogue to gain a better understanding of how to connect water and life.

Applying the best possible technology is only part of the answer. People and organization as well as business processes determine if a utility will be able to meet future challenges. Utilities have different scopes, legal frameworks, and purposes, which means that the term "business" process may also include organizational, utility, and service processes. A utility may be a combination of water business and steward of commons.

A ***smart*** water operation—which is the old paradigm—should satisfy the requirements of producing an acceptable product while keeping energy and resource requirements at a minimum. The new and higher goal is to integrate the complete urban water cycle. Still, many agencies don't have authority or scope of the whole urban water cycle, so partnerships, collaboration, or regional coordination is needed to achieve the goal. To tune the organization for external collaboration is certainly a difficult task.

It is great when technology can enable utilities to be smarter; indeed, technology can greatly enhance some of our capabilities. However, although technology has made significant advances, many tasks in utilities are still done by people. Gaining insight to issues related to the behaviors of individuals and organizations is, therefore, especially important. Furthermore, if people do not see that technology brings value to their work, or if they feel threatened by it, there is a great risk that the potential benefits of technology will not be realized.

A ***wise*** organization—the new paradigm—would include a broader set of capabilities that address the full scope of decision making including both humans and nature. It assumes active engagement, collaboration, and sharing by the participating utilities and the external users and environment. The WISE Framework has been developed to better describe how the overall enterprise works "in real life" and how different aspects of utilities interact with each other.

A wise organization would have the capability to:

- *Define—or at least articulate*—its purpose and values, adaptable to changing society or environment requirements.
- *Discern* how value is created and how it influences significant external entities.
- *Decide* based on knowledge acquisition and continuous and collaborative information collecting. The decisions have to adapt to new or additional information. Also, values are drivers for decision. Decision makers need mental flexibility, openness, and creativity on making effective judgments even when full certainty cannot be achieved. There is a clear path from data to knowledge and further to wisdom.
- *Deploy—and facilitate*—methods for effective learning and change and assessment of the value system.

The four capabilities interconnect in different ways that can be explained by the feedback principle of continuous action-and-reaction. Both technical and management decisions can be described by similar feedback structures. Our message is not that being smart is an unworthy goal for a utility; quite on the contrary—it is very desirable. We see the concept of a WISE utility as a more inclusive and integrated approach that facilitates different types of improvements, including a utility becoming "smart." Individuals and organizations must make their own determination if "smart" is sufficient for them. Organizational culture plays an important role in the decision-making process both in crises and during daily organizational life. There is an apparent relationship between the individual ethics of organizational leaders and the ensuing ethics of the organization itself.

Systems thinking is fundamental for the wise utility. It deals a lot with relationships and patterns, and how various "components" interact with each other. Sometimes we need a component view, and other times we need a systems perspective, like looking at objects from the ground or from a helicopter. The true challenge is to understand the component and process aspects from a systems perspective and to comprehend the system aspects from a component perspective. It all boils down to a more beautiful, meaningful, and comprehensive understanding of our role and our common work to add value to the overall water cycle.

Chapter 3 describes how business processes, workforce/people, and technology build up the value creation in a utility. Business process models encapsulate the knowledge about how an organization performs work and creates value. Facilitating integration is a key challenge to internal value creation. Each one of the value creation components can be assessed according to an appropriate maturity model. Such a model explains how well something is currently working in relation to its potential, how it *can* work.

A Water Sector Value Model (WSVM) has been developed, based on research results on organizational models and people behavior, further discussed in Chapter 5. It aims to provide a mechanism for utilities to share their business processes and methods. Business processes should be structured, and people should be informed, inspired, motivated, and educated, so that technology can be an enabler to create value. The role of technology for value creation is not currently in the scope of the UAIM project but will be further studied in the WISE research program.

People make up the crucial "human capital" of an organization. The individuals determine to a great extent how value is created. The "sense and purpose" has a huge power on the way we look at our tasks. It is crucial to structure business processes and to motivate and educate people so that technology can support the business processes to create value.

Integration is a key ambition. This becomes apparent when a utility is facing an "enterprise-wide" task that transcends the boundaries of a single organizational unit. A silo mindset cannot address challenges like climate change, extreme weather events, symbiosis establishment with surrounding industries, or advanced asset management. Lack of internal integration is often an obstacle to manage complex problems. Here we consider the integration of people, processes, and technology within the organization. However, integration also involves symbiosis together with external industries and surrounding water resources.

One remarkable outcome of the UAIM research revealed that the key challenges experienced by the utilities were dominated by "people issues" (46%) and business processes (43%) and only marginally by technology (11%). Challenges often occur when technology is introduced into an organization without considering the broader context including people, improved processes, and other factors. Consequently, utilities have to promote a higher maturity toward capability.

Chapter 4 considers how availability and effectiveness of water infrastructure influences every person's daily life. Utilities are traditionally *production*-oriented, but there is an emerging focus on customer- or *demand*-oriented operations. With intensifying water scarcity, the issue of *water reuse* becomes increasingly important. The opportunity to recycle water in big scale has not yet been adapted technically or mentally, and it will mean a huge transformation of the identity of a utility. It is a challenge for any utility to influence *water consumption* to avoid or delay investments in production capacity. The tariff (water rate) structure is a key challenge that should combine the goal that water is a human right with the requirement to run the operation of a water utility economically. The Maslow hierarchy model demonstrates that water is fundamental to every person. The model also applies to a water utility organization. How do we make the utility ready for major changes, caused by water scarcity and climate change?

Chapter 5 continues the discussion on qualities of a wise utility. Feedback from and dialogue with citizens (customers) and environment should be an input for organizational learning and change management. The reaction from external entities is more complex for a public utility compared to a private company and will include several nonmarket factors such as affordability, social justice, and stewardship of precious natural resources. Building a wise organization requires that we understand how humans make decisions. This then leads to how organizations make decisions. The chapter ends with a discussion about the road toward a wise utility. At planning, strategic, management, and policy levels, there are numerous complex challenges that have been characterized as "wicked" problems. Sometimes wicked problems are (almost) impossible to solve, which may be true with the old mindset. However, we may replace the term wicked with VUCA (volatile, uncertain, complex, ambiguous) problems to more correctly address the complex but solvable problems with a new mindset. A wicked problem involves many people and opinions, or economic obstacles. To handle such problems requires wisdom, which is a process of continuous learning, a journey rather than a destination. The chapter finishes with a practical issue that at a first glance does not appear to be wicked at all: managing assets in a utility.

Chapter 6 includes results from the first three years of the UAIM effort and describes different components of the WISE Framework. The key results are business processes improvements and people-side maturity and assessment models that help enable positive change. These are key components in the value creation and have been the center of the UAIM results. A brief overview is presented about the target of the WISE project and related research efforts. The documentation of new results will be updated via a web portal on a continuous basis. In this way, we aim to keep the document accessible, alive, and relevant.

Chapter 7 presents an overview of the future work on the concepts and ideas that were started with the UAIM project and describes how this work will continue under the WEF/WISE program. In Section 7.1, technology aspects are discussed. Cyber-security and data protection/privacy present great challenges, so the need for a new breed of digital workers is obvious. In Section 7.2, the main goal of the WEF/WISE program is outlined: *to develop and apply a holistic analysis framework and a methodology based on system modeling to help utilities improve maturity of their capabilities and implement change management focused on value and overall performance.*

Chapter 8 presents a new effort, Leaders for Emerging Applied Practices (LEAP), that would be conducted independently and concurrently with WISE. Both UAIM and WISE are applied research efforts focused on a structured methodology for improvement of water sector utilities. The approach is tailored to water sector utilities as they are today and considers traditions and constraints common to such organizations; the emphasis is on improvements that could be achieved in the relatively short term and have an incremental impact. The LEAP effort will address key global sustainability challenges that may require a more transformative approach that would reshape the dominant paradigms in water sector utilities. LEAP and WISE will be administrated separately. LEAP takes an explorative and piloting approach and should be financed completely separately from the WISE project. The aim is that any results of the LEAP project should be freely and openly available to all utilities.

# 1 The UAIM Framework Experiences

| Key Messages | • UAIM Framework introduces a holistic view of value creation within an organization.<br>• UAIM aims toward the development of "enterprise-wide" solutions that address the overall needs of the enterprise.<br>• UAIM has developed a practical methodology using a systemic approach to improve the management of water sector utilities. |
|---|---|

At the 2016 WEFTEC conference in the United States, a group of water sector utilities met in an all-day workshop to discuss the need for a collaborative effort to develop a methodology to document, analyze, and improve their performance. The methodology was to focus on internal value creation, and the first discussion produced a high-level view of the value chain for the water sector proposed by the utilities during this workshop as shown in Figure 1.1.

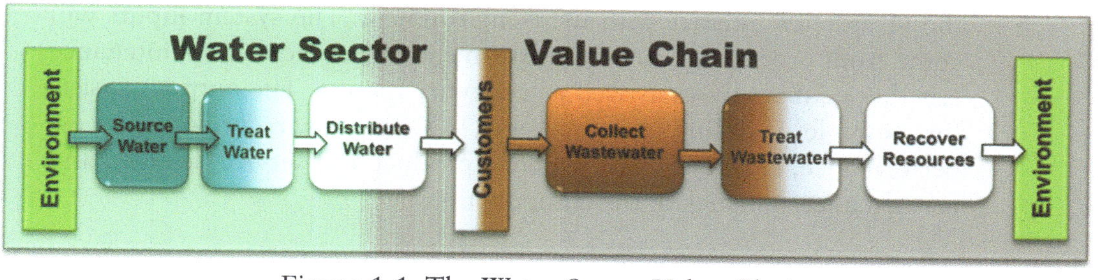


Figure 1.1. The Water Sector Value Chain

A conceptual model describing how value is created within an organization was introduced and accepted. This conceptual model, referred to as the UAIM Framework, was originally proposed by Cello Vitasovic. The goal was to provide a practical methodology that uses a systemic approach to improve different aspects of managing a water sector utility. The mission statement for UAIM is

> to develop and apply a holistic analysis framework and a methodology based on system modeling to help utilities improve maturity of their capabilities and implement change management focused on value and overall performance.

| *The goal of the UAIM Framework is to provide a practical methodology using a systemic approach to improve different aspects of managing a water sector utility.* |
|---|

Three concepts form the basis for the whole UAIM effort: systems, models, and feedback.

**Systems**

An organization can be described as a social system that interacts with the environment. A system is composed of components that work together to obtain a common goal. The people in the organization are the "components" of the system. The system is influenced from the outside—the external inputs, effects, or disturbances such as water availability, weather, regulatory rules, energy prices, customer input and feedback, and so on. The system will influence its surroundings—the system outputs—such as customer or citizen satisfaction, service levels, performance, cost, and nature.

There are several sub-systems within an organization system. The system components consist for example of computer hardware, instrumentation, software modules, and wastewater treatment process units. For example, the IT systems interact with their environment. The system inputs will come from process sensors and instruments, asset records, maintenance schedules, lab analyses as well as customer reports. The output consists for example of online measurements, lab analyses, records of process performance, customer requirements, product and service delivery, and performance statistics of the components.

An activated sludge process is another example of a system. The system components consist of biological reactors, hydraulic lines, pumps, compressors, and online instrumentation. The process is influenced by external inputs, such as influent flow rates and concentrations, and by manipulated inputs (control variables) like airflows and chemical dosages. The output from the system is characterized by operational data (online measurements and lab analyses) from the various processes, effluent quality, and energy efficiency.

**Models**

A model is a condensed summary of our knowledge of a system. A model can also be an operational support to complete a certain task. A model has always a purpose. Therefore, there may be several models of the same object. For example, many different models characterize a car. The driver has one model, the engine designer another one, and the body designer still another model.

A *descriptive* model of an organization provides insight in how the organization performs its functions. It is a systematic guideline or road map on how to achieve a certain task, often based on a specific event. The model can be defined in words or in block diagrams and does not have to involve any mathematical terms.

A *dynamical* model of a technical process (such as a pump, a biological reactor, a compressor) describes the behavior of this system as a function of time in response to manipulations or to disturbances. Such a model is typically described by mathematical equations, but can also appear as logical expressions like "if A happens, then perform B."

*Maturity* models present a basis for assessment of how well the functions of a system are performed, such as business processes, people, customer response, maintenance functions, and technical processes.

**The Feedback Principle**

We experience feedback all the time in our daily life. We may illustrate the concept by an everyday experience, driving a car. The driver is part of a feedback loop. The driver ensures that the car position and speed (the real process) are within given limits (discern, measure, observe). The eyes watch the instruments and the surrounding area and look for changes or "disturbances." They motivate that the speed or direction have to be corrected all the time. The brain analyzes the "measurement" data and decides how to change speed and direction (decide, control). The decision ("control signal") is transferred to muscles ("actuators") that will turn the steering wheel or the accelerator (deploy).

Receiving feedback and adjusting is an essential part of our decision making and learning. The "purpose" (the goal or the performance index) is the immediate goal to keep the car on the road but can also include the choice of destination; it provides the foundation for the driver's behavior and guides his decisions (define). Throughout this document, we will use the feedback concept and apply it within the context of utility management.

Sometimes the controller is automatic, as in most technical equipment or processes. At other times, the controller is a human being, as in management decisions. Defining a proper goal (performance index) is crucial, that is, assessing the consequences of the control action.

Depending on the chosen performance index, the resulting operation may vary dramatically. The human factor also comes into play; and trust, engagement, behaviors, cultural alignment with goals, and taking risk are important variables and potential barriers.

Norbert Wiener (1894–1964), the true pioneer of cybernetics (from Greek *kybernetes*, "steersman") defined cybernetics as the science of "control and communication in the animal and the machine." Early on in his work on cybernetics, Wiener was aware that feedback is an important concept for modeling not only living organisms but also social systems. He wrote in 1948: "It is certainly true and the social system is an organization like the individual, that is bound together by a system of communication, and that it has a dynamics in which circular processes of a feedback nature play in important role." Norbert Wiener's milestone publication[2] defines the basic principles of cybernetics that has later developed into control and communication.

The principles of feedback control are applicable all the way up to high-level strategic decisions. The framework is always the same, whereas the measurements, the analyses, and the decisions are different. Hence, it is vital to understand this way of thinking for utilities to become smarter, more robust, resilient, efficient, effective, and most importantly, more sustainable.

UAIM has established a framework for a *systemic approach* to value creation in a utility. Any system or sub-system in a utility is characterized by:

- the system boundaries,
- the components of the *system,* and
- the interactions between components.

The systemic approach provides an analytical "map" to examine the system behavior and performance. An integrated system has qualities that the sum of the individual components cannot explain. It is by a holistic view that the system—utility, community, and environment—can be learning and developing in a more sustainable way. The systems approach is not necessarily in conflict with the "component perspective" but offers a new way of handling key challenges. Systems thinking emphasizes connectedness, relationships, and context. The whole arise from the interactions and relationships between the parts.

> ***An integrated system has qualities that the sum of the individual components cannot explain.***

The UAIM Framework combined two familiar concepts:

- People, (business) processes, and technology as key generators of value;
- Strategic, tactical, and operational levels of utility management based on the time horizons for business processes, decisions, and operations.

By combining these two concepts, the UAIM Framework introduced a holistic view of value creation within an organization as shown in Figure 1.2.

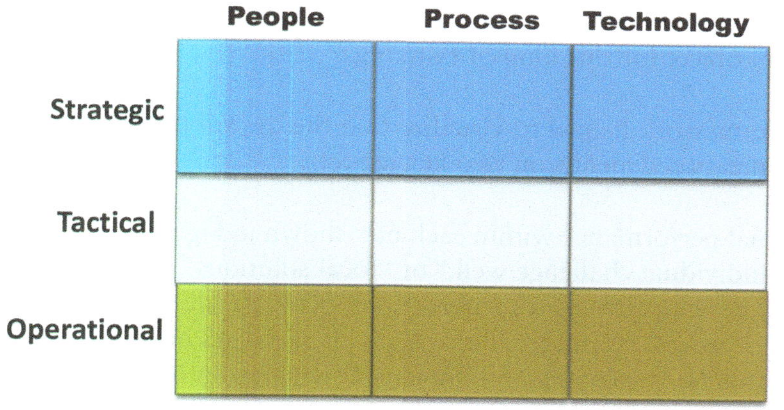


Figure 1.2. The UAIM Framework for Value Creation in an Organization

> ***The UAIM Framework is providing a holistic view of value creation within a utility.***

In early 2017, the UAIM Framework and the associated ideas were used to launch a research project with the WE&RF.[3] Several water sector utilities (referred to as the UAIM Utility Partners, see Appendix) committed to contributing significant cash and in-kind contributions required to fund this effort. The first workshop of the WE&RF research project was held at the 2017 Utility Management Conference.[4]

The main drivers for the development of this framework included:

- The desire and the need to understand better how the overall enterprise works "in real life" and how different aspects of utilities interact with each other: developing a *model of an enterprise* that could be used to improve our understanding and enhance the performance of a utility. The need for an integrated approach was recognized, leading to the development of enterprise solutions.

> In past decades, water resource utilities have developed a number of "local" solutions, such as technologies and IT systems that address different specific areas of their enterprise. However, the development of "enterprise-wide" solutions that address the overall needs of the enterprise, including work process documentation and improvement and people-side enhancements, is still lagging. There is a well-recognized difficulty to integrate different key software solutions—such as geographic information systems (GIS), Supervisory Control and Data Acquisition (SCADA), document handling, finance software, and so. The trend, however, is a further fractionating and budding—leading to increasing complexity and need for mastering many different systems. Alternatively, a single enterprise-wide system is not attractive since it would monopolize the system. It is essential to treat data as the primary value of the system and then develop systems around these data. Asset management is discussed below as a typical object for this kind of thinking.

The UAIM Framework helped to visualize that the overall (enterprise) performance of a utility organization depends on two key aspects:

- Optimal performance within each box shown in Figure 1.2: that is, "addressing each individual challenge well," or "local solutions."
- Effective management of interactions between all the boxes—integration of people, processes, and technologies in all parts and on all levels of the organization: that is, developing and implementing "enterprise solutions."

It may be helpful to distinguish between two different objectives or characteristics: *efficiency* and *resiliency*. They are slightly in competition, and the aim is to find just the right balance between them. There is a true struggle to be efficient in each box as well as between boxes. One has to find a good balance here encompassing both efficiency (no waste processes) and resilience (capacity to recover quickly from difficulties). A singular focus on efficiency may prohibit cross-organizational collaboration.

***Efficiency and resilience should be balanced.***

Figure 1.3 illustrates a UAIM higher-level summary view of internal value creation: people interact within the organization to execute business processes that are enabled by technology.

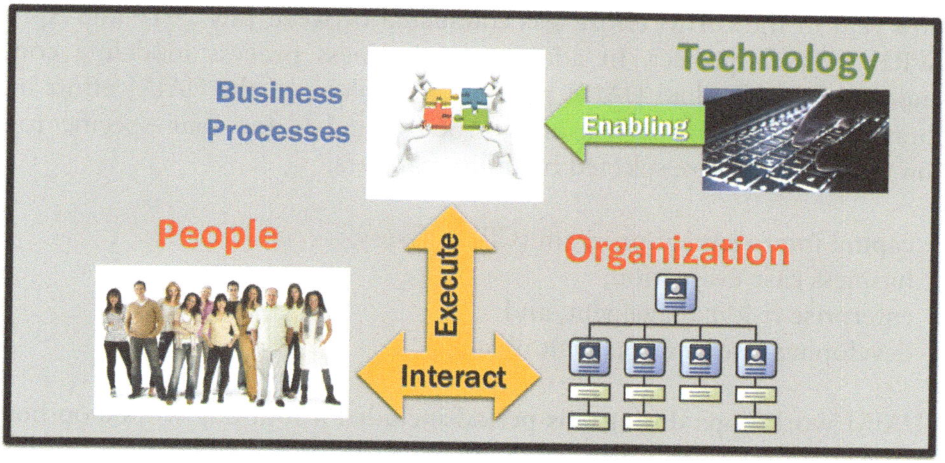


Figure 1.3. UAIM Internal Value Creation Model Within an Organization

The initial focus for the UAIM research project was on the business processes: the middle column of the UAIM Framework. A ***system model*** explains how something actually works. In particular, a ***business process model***[5] can answer a number of specific questions, such as how the process looks like, who is responsible, how the performance is assessed, how documentation is stored, and what decisions drive the process. More details of the business process modeling results are described in Chapters 3 and 6.

During the first year of the project, the utility partners adopted a standard method and notation for business process modeling and developed detailed business process models for areas that were of specific interest to their utility. This produced detailed and consistent documentation of some of their existing ("as is") business processes. All the business process models were posted to the UAIM knowledge base to be shared with all the UAIM utility partners. Workshops at the 2017 WEFTEC and the 2018 Utility Management Conference (UMC) were conducted to share and discuss the results of the first year of research.[6]

The second year of the research project was started after the 2018 UMC workshop and concluded with the 2019 UMC workshop. An additional workshop was conducted midyear at the 2018 WEFTEC. In the second year of the project, utility partners prepared case studies that documented their business process modeling efforts, and some case studies included the analysis of the existing "as is" business processes and the design of the improved ("to be") business processes. The results were shared and posted to the UAIM knowledge base, and the report is available from the WRF.

***Analyzing existing "as is" and designing improved "to be" business processes can build bridges of step-by-step improvements from point A to point B.***

The third year of the UAIM effort was conducted between July 2019 and April 2020 as a WRF research project. In addition to business process modeling conducted independently by individual UAIM partners, this phase of the UAIM effort included collaborative development of business process models for four specific topics of common interest that were selected by utility partners:

- capital improvement program (CIP) delivery,
- business case evaluation,
- enterprise risk management, and
- developing asset management plans.

Three UAIM workshops during this period included a four-day workshop hosted by the Clean Water Services in Oregon in September of 2019, a one-day workshop at the 2019 WEFTEC, and a two-day workshop at the 2020 UMC. In addition to workshops, this collaboration included numerous interactions and numerous conference calls by each of the four groups of utility partners. Common objectives, "future state" business processes, key challenges, and metrics were developed.

The UAIM as a WRF research project has been completed in 2021.[7] The UAIM effort will proceed in a different form as an industry initiative supported by WEF.

# 2 Water Intrapreneurs for Successful Enterprises—A Vision

| Key Messages | |
|---|---|
| | • Any utility needs to acquire a true understanding of the relationship between urban water systems and the surrounding nature.<br>• Climate change represents a truly challenging significant challenge of this century and poses risks to water and sanitation services. Climate monitoring is fundamental for every level of decision making in the city of the future.<br>• Applying the best possible technology is only part of the answer. People and organization as well as business processes and customer acceptance determine if a utility will be able to meet future challenges.<br>• A ***smart*** water operation should satisfy the requirements of producing an acceptable product while keeping energy and resource requirements at a minimum. The new and higher goal is to integrate the complete urban water cycle.<br>• A ***wise*** organization—the new paradigm—would include a broader set of capabilities that address the full scope of decision making, including both humans and nature. It assumes active engagement, collaboration, sharing, and learning by the participating utilities and the external users and environment.<br>• The WISE Framework has been developed to better describe how the overall enterprise works "in real life" and how different aspects of utilities interact with each other.<br>• *Descriptive models* provide insight in how an organization performs its functions. *Maturity models* explain how well something *can* work.<br>• The *assessment method* measures how well the system *actually* works. |

In this chapter, we aim to provide an overall vision, given current and future water sector challenges. A detailed description is given how value is created and how a utility must be subject to continuous learning. In Section 2.1, we discuss major challenges that should require the attention of the future utility. Essential capabilities of a wise utility are suggested in Section 2.2. We propose a way of assessing these qualities. The value of the key elements of a utility—people, business processes, and technology—should be given a performance indicator that we call *maturity level*. Similarly, external impacts of the utility and its operation create a value system that should be evaluated. Section 2.3 explains more details of the value creation. All the key components of value creation—people, processes, and technology—depend on each other.

## 2.1 Urban Water Challenges—Globally and Regionally

In his groundbreaking book from 1973, *Small Is Beautiful—A Study of Economics as if People Mattered*, the economist Ernst F. Schumacher said, "Wisdom demands a new orientation of science and technology toward the organic, the gentle, the elegant and beautiful."[8] Our modern societies have not experienced ourselves as part of nature, but as an outside force. We even talk about a battle with nature. However, if we win the battle, we would find ourselves on the losing side. This insight is crucial for a WISE utility. Climate change and the challenges of sustainable operations should govern the utility.

### 2.1.1 Nature and Water

The obvious fact that water is life should have a profound impact on how we extract, treat, use, and return water to nature. If water is in abundance, we seldom pay attention to any imbalances in nature. In our high-technology and urbanized lives, we often forget our connection to and dependence on nature. With increasing population and urbanization, higher demand from industries and food production, intensified by climate change and more erratic precipitation patterns, the balance between human water use and nature is threatened. "There needs to be some kind of water stewardship that ensures that the urban and the natural water cycles work together seamlessly and without destroying values in either place."[9] We need to acquire a better understanding of the relationship between urban water systems and the surrounding nature, not only in general or in global terms but in each specific location.

***A better understanding of the relationship between urban water systems and the surrounding nature must be acquired.***

Increasing water scarcity is a reality in many regions today, and the problems will be intensified. Water for municipal consumption is competing with industrial use and food production. Nature also depends on water and if our water operations and water use are not sustainable, then nature will suffer. In the end, this will have a damaging impact on us.

With increasing scarcity, we should establish incentives for the users to utilize water prudently and responsibly. An issue of increasing importance is how to design water tariffs to satisfy different needs. Tariffs should reflect not only production costs but also water availability and degradation. Even the poorest should afford the necessary amount of water and issues of cost; equity and environmental justice need to be integrated into the equation as well. This is further discussed in Section 4.2.

### 2.1.2 Climate Change

Water is a primary indicator of climate change. Climate change represents a true challenge of this century and poses risks to water and sanitation services. The threats relate to changes in temperature and precipitation, leading to changes in hydrology and water demand, as well as to storm events that damage water and power supplies. The nature of the threats relates to increasing unpredictability in surface water flows and a consequent change in demand for groundwater, as well as floods and declining water availability and quality. These changes may be experienced in the same location at different times.

> ***Climate change represents a truly significant challenge of this century and poses risks to water and sanitation services.***

Droughts and floods are becoming increasingly common. It may be possible to predict the areas that are vulnerable and to some extent when specific events may occur. Once underway, however, these events offer limited time and choices to take action. Long-term changes include sea-level rise, drought and water scarcity, changes in water quality, and flooding. Consequently, urban water utilities are facing an increasing need to improve the management of water resources emergency plans, and associated infrastructure. Diversifying sources of water supply will become increasingly important, through appropriate and sustainable extraction of groundwater, water trading or conservation, or use of recycled or desalinated water. Large utilities around the world are working with their perceptions, experiences, and approaches to addressing climate-related challenges of urban areas.[10]

In most places, planning still has been based on historic levels of water availability and consumption. Climate change, however, is causing important shifts in these patterns, having long-term economic, social, and environmental impacts. Planning for dealing with climate-related issues must be coordinated and based on sound principles of Integrated Water Resources Management (IWRM).[11] The most common exposure to climate change impacts reported among utilities are:

- decreased surface water quantity,
- increased urban demand for water,
- decreased surface water quality,
- increased competition for water resources, and
- decreased groundwater/aquifer recharge and quantity.

The most common actions taken by utilities to meet climate change have been:

- reduce consumption,
- monitor changes to improve watershed,
- reduce nonrevenue water (leakages), and
- recycle wastewater.

Climate monitoring and prediction is fundamental for every level of decision making. Water and sanitation services contribute to greenhouse gas emissions. Choice of wastewater treatment technologies, improved pumping efficiency, use of renewable sources of energy, and within-system generation of energy offer potential for reducing emissions. There is a significant potential to generate much of the energy requirements in a greenhouse gas neutral way from within the systems and potentially to be a net contributor to energy.

Naturally, there are water challenges where we have no solutions today. An interesting way of thinking and dealing with increased water scarcity and climate change is proposed in the "factory as a forest" discussions. A key challenge has been "if nature designed a company, how would it function?" This thinking has helped to create a framework to redesign business development to become one that had no negative impacts and a restorative influence. The circular economy should be an obvious component of the thinking.

### 2.1.3 Sustainability Goals

The 17 Sustainable Development Goals (SDG), defined by the United Nations, express what we need to achieve. Almost all of the SDGs will be influenced by utility actions, not only the SDG6—clean water and sanitation. Without clean water, it is challenging to fight against poverty (SDG1); eliminate hunger (SDG2); guarantee good health and well-being (SDG3); gender equality (SDG5); affordable and clean energy (SDG7); decent work and economic growth (SDG8); industry, innovation, and infrastructure (SDG9); build sustainable urban areas (SDG11); responsible consumption and production (SDG12); and climate action (SDG13). Water utilities have a direct impact both on life below water (SDG14) and on life on land (SDG15), and water plays a significant role in peace, justice, and strong institutions (SDG16). Without clean water and adequate sanitation combined with affordable energy (SDG7), hardly any of the 17 SDGs can be realized.

There is an opportunity to link with other industries to facilitate holistic delivery of the UN SDGs. For example, Thames Estuary Asset Management 2100 (TEAM 2100) program in the UK (https://www.gov.uk/government/news/thames-estuary-asset-management-2100-programme-team2100) is a 10-year program that links strategic objectives to asset management and a delivery process to refurbish, maintain, and replace tidal flood defenses in London and the Thames estuary. One aim is to establish partnerships with the banking industry in one of the policy units (flood risk cell) to deliver more than the obvious flood risk management outcomes. The SDGs that the program could deliver were mapped with some extra funding, and the local banking corporations were invited to participate in the initiative. This demonstrates an enhanced opportunity for wider collaboration.

In general, water utilities need to reach out as to SDG 17 partnerships for the goals and establish industrial symbioses, collaboration with customers in water savings, as well as collaboration with nongovernmental organizations.

### 2.1.4 The WISE Approach

Applying the best possible technology is only part of the answer. This publication aims to show that people—utility staff and users—and organizations as well as business processes determine if a utility will be able to meet future challenges. This will require *wisdom*. Next we will try to explain what a **water wise utility** may look like. Our goal is to encourage an ongoing dialogue to achieve a better understanding of the utility of the future.

> ***People—utility staff and users—and organizations as well as business processes determine if a utility will be able to meet future challenges.***

## 2.2 Qualities of a Wise Utility

When applied to organizations, such as cities or water sector utilities, the word "smart" is typically associated with using advanced technology intelligently and diligently.[12] Smart utilities and intelligent water systems that leverage technology can indeed provide important new capabilities to organizations.[13]

A smart water operation—the old paradigm—should satisfy the requirements of producing an acceptable product while keeping energy and resource requirements at a minimum. Such a system should respond quickly and adequately to disturbances and recover rapidly after a major upset. The operation of the system must be sufficiently transparent so that people involved in the operation have enough information to make informed rational decisions.

These qualities are also supported by the notion that utilities are managing capital-intensive assets and operations. The development of asset management capabilities are key, and such assets should be managed to achieve the short-term and long-term sustainability of assets. Moreover, the utility should produce desired levels of service, at minimum cost, consistent with acceptable levels of risk.

Organizations (utilities) can use technology to add value in two major ways:

- *Automation*: implementation of technologies to make automated decisions based on collected data and analytics.
- *Decision support*: technology is used to collect, analyze, and provide information to human decision makers and thus enable and improve their decision making. Already on the operational level, this requires that people are sufficiently skilled to correctly make use of the information. However, smart technology is not sufficient at higher-level decisions. Reality is considerably more complex than meeting effluent requirements and saving resources. It involves stakeholders, economy, technology, politics, environment, and social dynamics, all mixed and tangled.

***A smart water operation produces an acceptable product while keeping energy and resource requirements at a minimum. People involved in the operation have sufficient information to make informed rational decisions.***

The goal of a smart system is to integrate the complete urban water cycle: water supply, water distribution, urban drainage, wastewater treatment as well as the customer or demand side. Still, many agencies don't have authority or scope of the whole urban water cycle, so partnerships, collaboration, or regional coordination is needed to achieve the goal. To tune the organization for external collaboration is certainly a difficult task.

In this publication, the focus is extended from technology and automation to include the processes and decisions in which humans play a significant role, and technology acts as an enabler that provides support. Most important decisions in water sector utilities are still made by humans.

The variety and complexities of decisions that have not yet been automated or enhanced is considerable. To address challenges where technology is only in a supporting role, we need a concept that fully considers the role of humans, as individuals and as organizations. We may reflect what it would mean that an organization has a social, a cultural, and a collaborative character.

We would propose that the word "wise" better describes the attributes of a well-managed organization.[14]

This expanded concept from "smart" to a "wise" organization would include a broader set of capabilities that address the full scope of decision making that includes humans. A wise organization would have the capability to:

- **Define**—or at least articulate—its purpose and values, adaptable to changing society or environment requirements.
- **Discern** how value is created and how it influences significant external entities.
- **Decide** based on knowledge acquisition and continuous and collaborative information collecting. The decisions have to adapt to new or additional information. Also, values are drivers for decision. Decision makers need mental flexibility, openness, and creativity on making effective judgments even when full certainty cannot be achieved. There is a clear path from data to knowledge and further to wisdom.
- **Deploy**—and facilitate—methods for effective learning and change and assessment of the value system.

Methods based on these four capability principles will be applied to assess qualities of various features within a utility and of its external impact.

There are many benefits for a utility becoming "smart." However, acquiring wisdom is a long-term journey, including learning and transforming, that offers additional values. There are typically several learning steps between "smart" and "wise." Any individual or organization must determine if "smart" is sufficient for them, and how people, technology, and processes are to be integrated.

A wise utility should be based on a value system and culture that

- recognizes the value of water,
- understands environmental and ecological issues,
- effectively engages the people who work in the utility, and
- respects the customers and the community.

> ***A wise organization would include a broader set of capabilities that address the full scope of decision making that includes both humans and nature.***

The obvious goal of any utility is to supply clean water and provide wastewater services to customers. An important step toward "wise" is to adapt to, protect, and preserve the natural water systems that we depend on and operate within. Ultimately, our human water consumption should have no detrimental effects on nature.

Both a smart and a wise utility are based on systems thinking, but with different mindsets. This will be examined further in the next section. Then we will examine each of the four key capabilities ("the four Ds") of a wise water sector utility, as defined, and will then consider the key capabilities of organizational wisdom within a framework that helps organizations assess and improve different components of success.

## 2.3 The Role of Systems Thinking

It is becoming more and more evident that major problems for a utility of today, as remarked in previous sections—climate change, water scarcity, environment, social problems, and financial challenges[15]—cannot be understood and solved in isolation. They are systemic problems, which means that they are interconnected and dependent. There are certainly solutions to many of these complex problems, but they require a radical shift in our perceptions, our thinking, and our values. There is a formidable challenge that leaders and decision makers at all levels should see that many of the challenges are interrelated. We have to "connect the dots." Furthermore, our solutions of today will influence future generations.

Systems thinking deals a lot with relationships and patterns, and how various "components" interact with each other. Therefore, we cannot deal with technology, people and their priorities and ambitions, and organizational issues as independent issues. All of them are interconnected. The concepts of "smart" and "wise" are not in opposition. Rather, a wise organization may mature out from a smart one. Sometimes we need a component view, and other times we need a systems perspective, like looking at objects from the ground or from a helicopter. The true challenge is to understand the component and process aspects from a systems perspective and to comprehend the system aspects from a component perspective.

Systems thinking is the opposite of analytical thinking. Analysis means *taking something apart* to understand it; systems thinking means *putting it into the context of a larger whole*. Therefore, we call systems thinking *contextual*.

***Component and process aspects have to be considered from a systems perspective, and the system aspects have to be comprehended from a component perspective.***

A *smart* mindset is more mechanical, whereas a *wise* mindset widens the perspective from objects to people and further to life itself. We may talk about the smart mindset as a *machine view*, and the wise one as a *living systems view*, as expressed in Table 2.1.

Table 2.1. The "Smart" and the "Wise" Mindsets (adapted from Capra & Luisi, 2019)

| **Current Paradigm—Machine View ("Smart")** | **Systems Paradigm—Living Systems View ("Wise")** |
|---|---|
| Fractionating, component focus | Everything is connected |
| Simplification | Complexity |
| Objects and affect | Structure and relationships |
| The whole is a sum of the parts | The whole is more than the sum of the parts |
| The parts determine the working of the whole | The whole determines the working of the parts |
| Context irrelevant | Context is key |
| The world as material | The world alive |
| World is a machine, and the universe is a machine | World as a network, the universe is organic, living, spiritual |
| Islands of explanations | Full explanation is impossible |
| Mechanistic | Ecological perception of the world |
| Shallow ecology | Deep ecology |
| Valuing primarily human design | Delving into nature's design |
| Nature has instrumental value | All life is valuable in itself |
| Human-centered value | Earth-centered value |
| The study of objects | The study of life |
| Focus on right answers | Continued questioning |
| World is objects obeying laws | World is an ordered and harmonious structure |
| Precision | Patterns |
| Finite | Infinite |
| Treating disease | Healing the system or balancing the system |
| Entropy (gradual decline into disorder) | Homeostasis (the condition of optimal functioning for the organism) and autopoiesis (a system capable of reproducing and maintaining itself) |
| Primary focus on quantity | Primary focus on quality |
| Linear[a] | Nonlinear[a] |
| Utilitarian purpose | Purpose of understanding |
| Standardization | Diversity |
| Material systems | Living systems |
| Anthropocentric (man in the center) | Humans as one strand of the web of life |
| Hierarchy | Network |

| | |
|---|---|
| Reductionistic[b] | Synthesis[c] |
| Mental | Whole person view |
| One-to-one rules | Symbiotic fit (mutually beneficial relationship between different people or groups) |

[a]An important property of nonlinear systems, which has been very disturbing to scientists, is that exact prediction is often impossible, even though the equations may be strictly deterministic (think of weather prediction). This will bring a shift from quantitative to qualitative analysis. While behavior in the physical domain is governed by the "laws of nature," behavior in the social domain is governed by rules generated by the social system itself.

[b]This implies that one should be able to understand all aspects of complex structures—nature, complex technical systems, or the human body—by reducing them to their smallest constituent parts. This philosophical position is known as Cartesian reductionism.

[c]The essential properties of a living system are emergent properties—properties that are not found in any of the parts but emerge at the level of the system as a whole. These emergent properties arise from specific patterns of organization—that is, from configurations of ordered relationships among the parts. This is the central insight of the systems view of life.

As expressed by Ingildsen (2020), "Something more than the 'ultimate good control of water' should be defined: looking for something called 'water stewardship.' 'Smart' or 'intelligence' must be supplemented by something of a different dimension, such as 'the best of humanity,' 'a caring respect for nature,' and 'poetic beauty.'" The following explanation of wisdom should define the ultimate goal of any individual or organization (Ingildsen, 2020): "To me, the difference between knowledge and wisdom is that wisdom is eternal and shows us a way of being in life—while knowledge is finite, being changed all the time. Knowledge has to be rewritten all the time—while wisdom is our sounding board for securing that our decision is based on a natural ground. Knowledge without wisdom is like water in the sand. Going beyond the famous Einstein quote, that you cannot change anything by using the same way of thinking, will demand of us to include our eternal wisdom coming from the heart."

***Knowledge has to be rewritten all the time—wisdom is our sounding board for securing that our decision is based on a natural ground.***

It all boils down to a more beautiful, meaningful, and comprehensive understanding of our role and our common work to add value to the overall water cycle. Nature is part of the water cycle, and this fact can have a profound impact on water operations. The theory of giving rights to nature was proposed in the 1970s by the American legal scholar Christopher D. Stone as a strategic environmental defense strategy.[16] Today, New Zealand's Whanganui River is a person under domestic law, and India's Ganges River was recently granted human rights. Bolivia and Ecuador have passed laws granting all nature equal rights with humans. In Ecuador, the Constitution enshrines nature's "right to integral respect." In practice, that means that all persons, communities, peoples, and nations can demand that Ecuadorian authorities enforce the rights of nature. One of those rights is the right to be restored. "Can nature trust us?"

> ***Trust is a fundamental asset of any organization.***
> ***Can nature trust us?***

Trust is a key resource of any personal or professional relationship, and it is critically important for a wise utility to develop and maintain trust between individuals and groups within the organization, between the people (workforce) and management, between the different external entities and the utility. Trust is a fundamental asset of any organization and the foundation for effective collaboration.[17] It starts with inclusion, as without that there isn't the opportunity to develop trust. Many times there is a desire on the part of decision makers to not include other diverse voices, internal or external, as they assume it will make it harder to make a "good" decision or to control the decision. It may become a very complex process with lots of conflicting goals and interests, where technology is only one part. Utility management can learn from social sciences, find good inclusive processes, and educate facilitators to perform these processes so that all stakeholders are heard, and decisions are made transparently. This is certainly not a straightforward process, but requires the acquisition of new capabilities, perhaps "wisdom."

## 2.4 The WISE Framework

The key capabilities of a wise utility are applicable to many aspects of utility management: that is, there are many things that we should do wisely. We will introduce concepts that connects key capabilities with the WISE Framework. The WISE Framework takes into account the internal value creation, the interactions with external entities, as well as organizational learning and change management.

***Descriptive models*** provide insight in how an organization performs its functions, while ***maturity models*** explain how well something *can* work (see the box on *models*, Chapter 1). The ***assessment method*** measures how well the system *actually* works. The development of the WISE Framework is driven by the desire and the need to better understand how the overall enterprise works "in real life" and how different aspects of utilities interact with each other. Each aspect of performance included in the WISE Framework can apply methods based on key capabilities as summarized in Table 2.2.

Table 2.2. Key Capabilities of WISE Organization

| Capability | How It Fits within the WISE Framework |
|---|---|
| **Define** its purpose and values, adaptable to changing society or environment requirements. | Organizational values drive the behaviors and determine how successfully an organization can improve internally and adapt to external challenges. Purpose and values should be translated to strategic decisions as well as to daily operations. |
| **Discern** how value is created and how it influences significant external entities. | An organization needs to document and fully understand how it creates value using processes, people, and technology to produce outputs that benefit external entities. Organization needs to collect and analyze data to produce information that enhances decision making. |
| **Decide** based on knowledge acquisition and continuous collaborative information collecting. | Decisions have to adapt to new or additional information. Values are drivers for decision. Decision makers need mental flexibility, openness, and creativity on making effective judgments even when full certainty cannot be achieved. |
| **Deploy** methods for effective learning and change and assessment of the value system. | Learning and change management are required in a dynamic world that is continuously changing around us. The obvious task to execute decisions should not be overlooked! |

***The WISE Framework aims to describe how the overall enterprise works "in real life," and how different aspects of utilities interact with each other.***

The four capabilities in Table 2.2 interconnect in different ways that can be explained by a feedback loop[18] of continuous action-and-reaction (Figure 2.1). Both technical and management decisions can be described by similar feedback structures. The "define" is called the goal or performance index in technical systems. "Discern" includes the collection and analysis of relevant data to produce information needed for decision making. Corresponding technology terms may be called "measurements, observations, analysis." "Decide" is the control action. Both in technical and managerial systems, decisions most often have to consider uncertainty. Finally, the decision created by a computer algorithm or in a boardroom has to turn into action, from "brain" to "muscles." In technical systems, motors, valves, pumps, or compressors are typical actuators. In management, the actions have a different character. In any system there is an external influence, sometimes as unwanted disturbances of the load into a treatment plant, at other times reactions from customers, from environment, or from significant weather events.

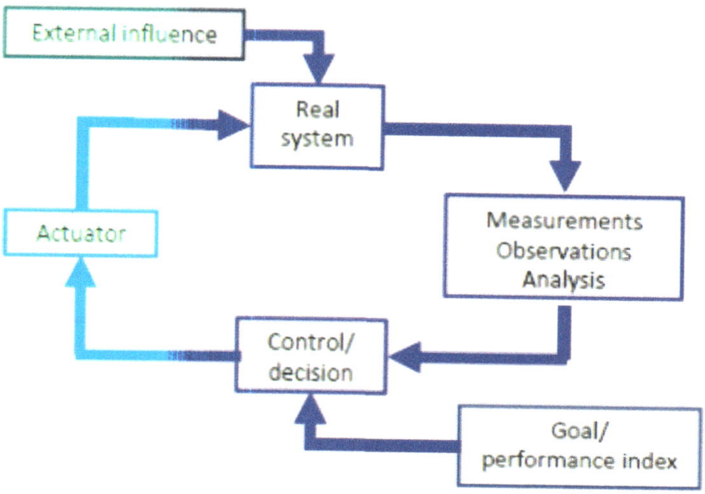


Figure 2.1. The Feedback Principle

Organization's capabilities are based on the level of knowledge that the organization has about different aspects of its performance. Two types of models are applied to encapsulate knowledge and provide a connection between the key capabilities and the overall WISE Framework (box on *models*, Chapter 1):

- *Descriptive models* provide insight in the underlying mechanisms and components. The models include business process models describing workflows, data models that explain how data are managed, and conceptual models that describe the employee engagement or an organization's culture and governance. These models provide an organization with a better understanding of how it performs different functions.
- *Maturity models* provide a basis for assessment of how well these functions are performed; the measured assessment of a specific capability is placed within a range that includes different levels of effectiveness. In addition to providing information about the current ("as is") state of a specific capability, maturity models provide clues about areas that can be improved and offer guidance for improvement ("to be"); in addition to value creation, maturity models may also be used to assess the quality of impact on external entities.

***Descriptive models provide insight in underlying mechanisms and components. Maturity models provide help to assess how well these functions are performed.***

Both types of models can help us understand, manage, and improve different aspects of internal performance and external impact for each of the components of the WISE Framework. Having the system models explaining how specific systems work will also help to increase maturity.

The UAIM Framework (as shown in Figures 1.1 and 1.3) focused on creation of internal value within a water sector utility. The framework has now widened into a broader context and is part of the WISE Framework that includes interactions with external entities, organizational learning, and change management as shown in Figure 2.2.

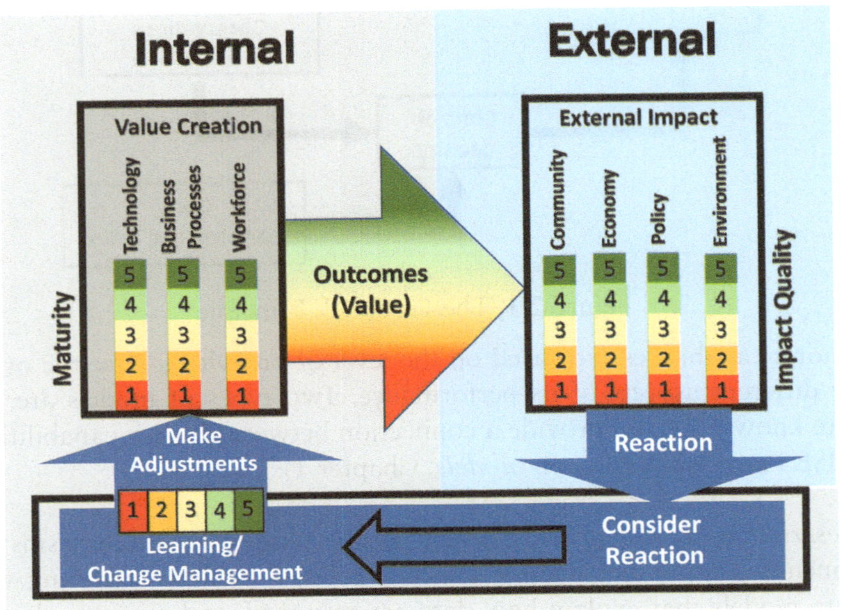


Figure 2.2. The WISE Framework—Single-Loop Learning

The upper left box in Figure 2.2 symbolizes how combined efforts of workforce (people, organization), business processes, and technology create value, not only in terms of money. The maturity is an assessment (on a scale 1–5) of how well these functions are performed. The most pessimistic case (level 1) is if the utility is blind and deaf to any signal from the external entities or from internally generated information that indicates the status and condition of its infrastructure. The utility operations have an impact on customers (community) as well as on economy, policy, and environment. The utility impact can be assessed (more or less apparent) using a similar scale 1–5, reflecting both economic value, customer satisfaction, and nature protection. All of these produce a reaction to the utility. As the utility considers the reactions, the feedback loop is closed. However, the quality of the adjustments made should be assessed, and adjustment should influence the internal operations.

Obviously, the result of the maturity evaluation depends critically on the evaluator. We are hardly able to make a trustworthy evaluation of ourselves. Instead, the assessment should be done by somebody else using more objective methods and scales. Similarly, an organization may have difficulties to detect its own shortcomings. Somebody outside the organization has a better chance to assess. Chapter 5 involves a deeper discussion on learning, motivations, and organizational culture.

Referring to Table 2.2, the four capabilities are illustrated in Figure 2.2:

- *Define purpose and values*: the value created within the organization, as shown in the top left of the figure, should be considered based on the outcome that it delivers to the external entities in the urban water system including the customers/citizens, the community, and the environment. Each of these value-generating components can be assessed based on the appropriate maturity models.
- *Discern internal and external challenges and impacts:* the outcome from internal value generation consists not only of easily measurable parameters of water quantity and quality: it includes impact on customers, community, economy, and affordability as well as asset and delivery reliability. Environmental impact of the water operations is essential and should be evaluated. External entities react to the value that they receive from the utility and provide a reaction back to the utility that reflects the difference between the value they have received, and the value they had expected or desired.
- *Decide based on data:* the utility receives the external feedback (reaction) and uses its learning (and change) capabilities to consider what type of adjustments may be required for processes, technologies, and people/workforce.
- *Determine path for learning and change*: after deciding what needs to be addressed, the utility determines how to implement adjustments to the internal value creation components (people, process, technology) to address the areas of concern.

Is there an ability or willingness to learn and adapt based on the external reactions? Will adjustments take place, based on the internal and external assessments? Some external reactions may be less apparent through the traditional lens and require more proactive attention and more complex assessment methods, such as social impact and environmental consequences. This kind of information is usually available from the records of a customer call center system. Fear of change can be a "showstopper." Therefore, developing knowledge around the reasons for change can mitigate the factor of fear.

The reaction from external entities is a feedback signal that primarily depends on two factors:

- How far is the actual outcome from the expected (or ideal) outcome? One way to quantify this would be the maturity level in the external box of Figure 2.2.

- The reaction also depends on the impact magnitude, probably measured as a cost.

The governance and culture will strongly influence an organization's ability to learn and change. Levels of maturity and methods for self-assessment of learning/change capabilities are a topic of current collaborative research by several water sector utilities within the UAIM project.

| ***The governance and organizational culture will strongly influence an organization's ability to learn and change.*** |
|---|

The governance and organizational culture provide the fundamental assumptions that drive behaviors and influence an organization's ability to learn and change, as illustrated by the double-loop learning in Figure 2.3.[19] When following this model of learning, we are willing and ready to examine and possibly challenge the organization's culture and the underlying system of the organizational values that drive the organization's behavior. Developing trust and new norms and behavior expectations can often be critical to effective change.

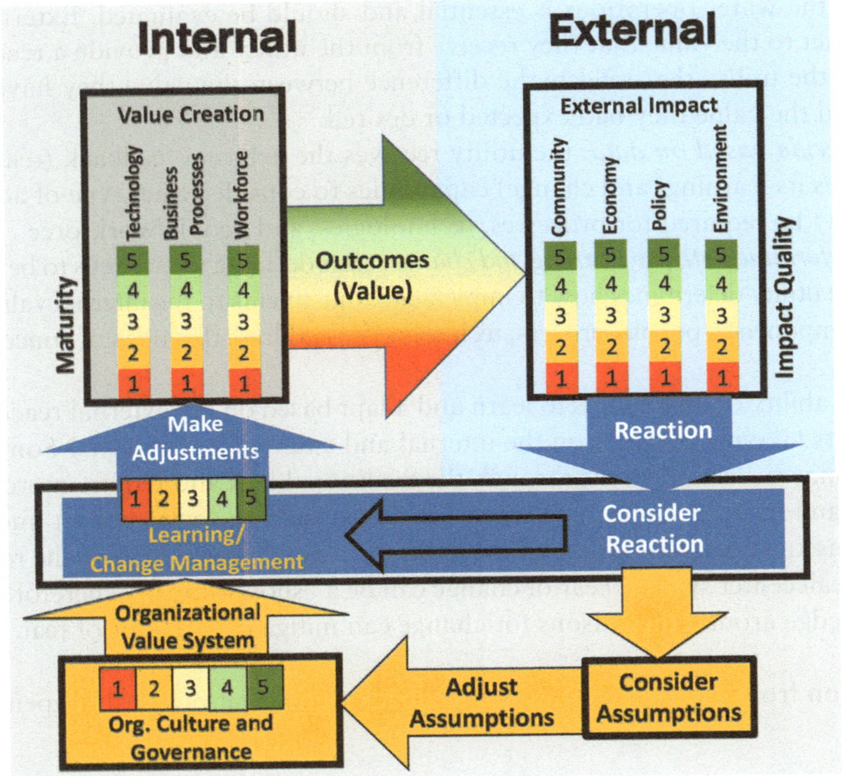


Figure 2.3. The WISE Framework, Including Double-Loop Learning
*Note*. The double loop is further examined in Chapter 5.

Figure 2.3 emphasizes that consideration of *organizational culture*[20] is of paramount importance. Organizational culture is a system of shared assumptions, values, and beliefs that governs how people behave in organizations. The organizational culture affects the way people and groups interact with each other, with clients, and with stakeholders. Furthermore, it may affect how much employees identify with an organization. Culture includes the organization's vision, values, norms, systems, symbols, language, assumptions, environment, location, beliefs, and habits.

Organizational culture plays an important role in the decision-making process both in crises and during daily organizational life. There is an apparent relationship between the individual ethics of organizational leaders and the ensuing ethics of the organization itself. There may be conflicts between new social values of employees and more traditional values of the organization. Cross-cultural human relationships may be complex and challenging and have to be managed. In short, a culture of collaboration is fundamental, but collaboration does not mean that everybody has the same opinion. Reciprocity is the glue that governs the myriads of social exchanges that take place every day in an organization, most of which have nothing to do with the realm of formal management. It is vital that leaders build transparent and open communications and incentives to reward fair, equal, reciprocal, and resilient exchanges between people.

Individuals need to be comfortable with challenges from others, from the environment, and from changing policies. It means that people are comfortable with being challenged and to challenge in a respectful way. This can be a disruption to traditional, hierarchy relationships, roles, and responsibilities, and require intervention and new models, systems, organization, or resources to maximize effectiveness.

***Organizational culture is of paramount importance.***
***Individuals need to be comfortable with challenges from others, from the environment, and from changing policies. A culture of collaboration is fundamental.***

# 3 Value Creation

| | |
|---|---|
| **Key Messages** | • People, business processes, and technology make up the value creation components in a utility.<br>• Each one of the value creation components can be assessed according to the appropriate capability maturity model.<br>• *Facilitating integration* is key to internal value creation.<br>• The WSVM aims to provide a mechanism for utilities to share their business processes and methods.<br>• The way human capacity is used is crucial for the value creation in an organization. Issues related to people/workforce will be further examined during the next year by the WISE project initiative.<br>• Business processes should be structured, and people should be informed, inspired, motivated, and educated so that technology can be an enabler to create value. The role of technology for value creation is not currently in the scope of the UAIM project but will be further studied in the WISE research program.<br>• Technology hurdles are not only related to technology itself but occur when technology is introduced into a broader context of an organization that includes people and processes as well as external stakeholders.<br>• A silo mindset cannot address challenges like climate change, extreme weather events, symbiosis establishment with surrounding industries, or advanced asset management.<br>• Implementing effective cybersecurity will be increasingly challenging.<br>• With different organizations often being responsible for the water cycle, cross-agency collaboration will be required. |

The original UAIM Framework (Figure 1.2) provided a "map" to start addressing the challenges or internal value creation in a holistic way.[21] While this framework has offered valuable guidance for analysis of internal value creation, it became apparent that this framework would need to be extended. We note that the framework is composed of nine parts, which is correct. At the same time, this is the fallacy of reductionism, noted in Table 2.1. We need to understand the behavior of the complete utility, and this cannot be explained in terms of the individual boxes only. We also need to understand the couplings, relations, and connections between the boxes. This is the motivation for synthesis (Table 2.1) and systems thinking.

Using the systems approach, the original model (Figure 3.1 left) is included as a component ("Internal Value Creation") shown to the right and also located in the upper left corner of the more comprehensive WISE Framework shown in Figures 2.2 and 2.3. In Section 2, we also introduced the concept of maturity models added to each one of the value creation activities.

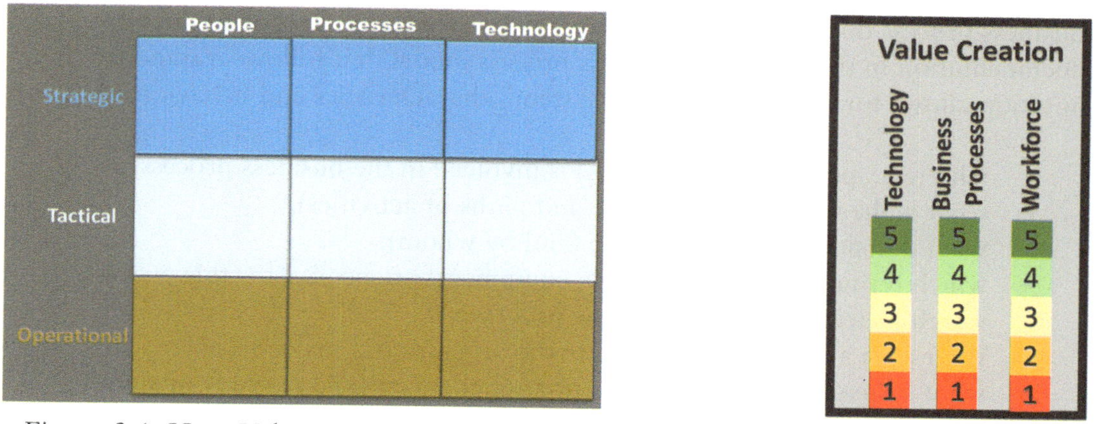


Figure 3.1. How Value Creation Should Be Added to Obtain the WISE Framework

> ***The original UAIM model has been integrated into the WISE Framework.***

Creation of value within a utility includes execution of different business processes and making different decisions by different people in different business units across the organization, using different technologies and practices. The most prevalent organizational structure in water sector utilities follows the traditional approach to the problem: organization is divided into units that provide different types of specialized work. This hierarchical approach to problem solving is frequent for organizations of many types: a utility needs people with different skill sets to work in accounting, engineering, operations, maintenance, laboratory, IT, and so on. Although it facilitates execution of specialized tasks, this approach also produces fragmentation that inhibits an organization's ability to solve problems that cross the organizational boundaries. Thus, a key to internal value creation is ***facilitating integration.***

> ***Fragmentation inhibits an organization's ability to solve problems that cross the organizational boundaries.***

We highlight the difficulties that fragmentation causes related to business processes (Section 3.1), people (Section 3.2), and technology (Section 3.3). In all cases, fragmentation causes hurdles in managing complex problems effectively. This illustrates the misconception of reductionism, as remarked previously. We need to understand how all these activities connect and depend on each other to manage a complex organization and operation. Consequently we consider integration, Section 3.4.

## 3.1 Value Creation—Business Processes

Business process documentation is lacking in many utilities. Producing such documentation in the form of a business process model that follows a standard notation allows for defined details of "system" characteristics and behaviors such as

- who (e.g., people, business units) is involved in the business process,
- what is the workflow (sequence and paths of activities),
- what decisions need to be made (and by whom),
- which resources (e.g., time, staff, equipment) are needed for this process,
- which metrics are affected by the process,
- what data are required to execute the process (and make decisions), and
- what system will be required to provide the data (and possibly analytics).

The business process is like a butterfly: one wing is the humans/workforce, the other is the technology (or the entire infrastructure), and in between the body that makes the two wings flap as a butterfly is the business processes. Most utilities have access to similar skill sets in people and similar technologies. The great difference is what they do with "the wings." What each organization does is codified in all the running processes—of which especially the business processes can be subject to a more or less smart/wise design. This can transform the utility to a smarter/wiser system as it makes us an increasing part of the full capacity in the joint system. Today there is so much capacity that goes unused or even undetected or hidden.

We will now describe the the WSVM. The goal of the WSVM is not to establish a single "optimal" or the "best" business process model that should be applied by all. The goal is to provide a mechanism for utilities to share their own business processes and methods across the spectrum of water utility functions.

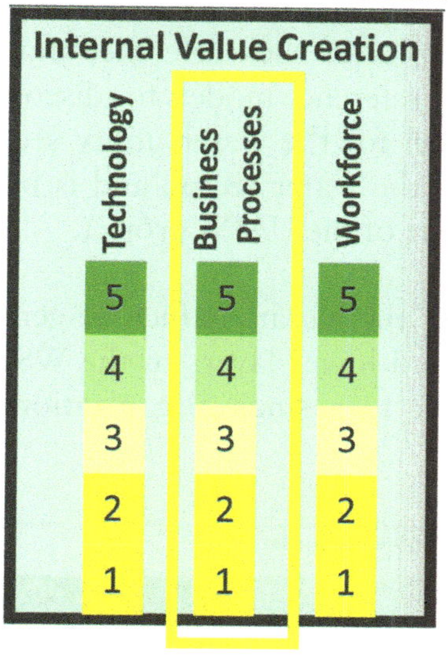


From Figure 3.1

Managing a water sector utility includes many business processes, executed by different units across the organization. This is how utilities get their work done.

Development of business process models was the focus in the early stages of the UAIM project by a number of participating water sector utilities. All the individual/specific business process models developed by the participating utilities were integrated into the WSVM. Under WISE the curation of the WSVM will be moderated by WEF strategic and technical committees.

Water sector utilities have been using models based on engineering and science knowledge for many years in terms of mathematical models of technical systems and processes.[22] They include hydrology in the collection systems, hydraulics for the pipe networks, computational fluid dynamics in clarifiers, aerobic and anoxic biological reactors, and dissolved oxygen transfer in activated sludge plants. These models encapsulate our knowledge and enable us to improve our decisions in both design and operation by:

a. defining the boundary of the system,
b. identifying the system components,
c. describing the behavior of components,
d. considering the interaction between the components,
e. identifying what can be manipulated, and
f. predicting the behavior of the overall system.

Business process models perform a similar function: they encapsulate the knowledge about how an organization performs work and creates value. Similarly to models based on scientific/engineering concepts, properly developed business process models also provide the capability to simulate and evaluate different "what if" scenarios to answer specific questions and inform decision making. To more completely describe a utility organization and management, we need to learn from social sciences and how this social system interacts with its environment.

***Business process models encapsulate the knowledge about how an organization performs work and creates value.***

Private industry has developed several "reference models" that describe the business processes that are common across the industry; examples of such reference models include supply chain for manufacturing, and e-discovery reference models for discovery processes in litigation. The WSVM is a reference model for the water utility sector: it includes business processes that are common across the water sector, and is being developed collaboratively by water sector utilities as part of the UAIM project.

The WSVM is organized hierarchically; upper layers of the hierarchy include generic high-level processes that are common to all utilities, and lower layers of the WSVM contain increasing levels of detail that may be specific to a single organization, as illustrated in Figure 3.2.

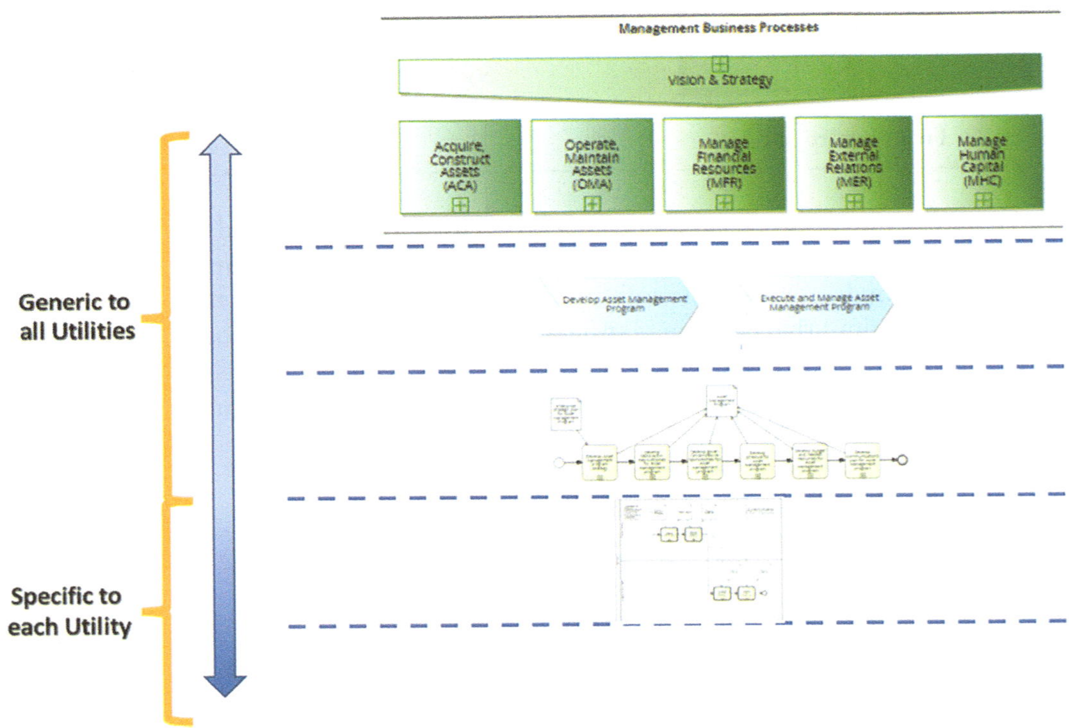


Figure 3.2. The Water Sector Value Model

Business process models developed by the UAIM utility partners during the past three years of the UAIM project have been posted to a shared "modeling knowledge base," and they also reside in a shared business process modeling environment, where they are all part of a comprehensive overall "model of a utility." The overall WSVM is far from complete: there are many business processes that have not yet been modeled. New business process models will be added as they are developed: the long-term goal is to "fill in the blanks" in the WSVM in the following years of the research effort. For some business processes, there will be alternative business models developed by more than one utility partner, with tailoring to their own organization. In the modeling effort, it is obvious that we should take advantage of all knowledge in each professional speciality.

There is no single unique model that fits everybody. Instead, the model is a platform for sharing business process experiences between utilities. If a person in utility A is looking to improve their own business process (e.g., procurement, or inspection of pump stations), the individual could go to WSVM and find what has been posted there by peer utilities. The individual could find a business process model that has been posted by utility B and learn how utility B is managing this process. The individual may find that the posted model has some good points, but it may not completely fit the needs of the utility since some of the conditions at utility B were very different (e.g., Minnesota in the north vs. Florida in the south). One could start with the model from utility B, modify it to fit the needs of the utility, and post an alternative model to the WSVM. Additionally, one could contact peers in utilities that had posted different versions of these processes and start communicating with them within the UAIM collaborative peer-to-peer platform.

Experiences and results of business process modeling and the WSVM are presented in more detail in Section 6.1.

## 3.2 Value Creation—People

When speaking of "workforce," we are talking about people: the employees or the "human capital" of an organization. Our focus is on individuals who perform much of the work required for organizations to create value.

From Figure 3.1

In business process models, people may appear in "swim lane" diagrams: we can see Jane in accounting and Bill in engineering project management as they have their places and roles within business processes. However, business process models do not capture if Jane and Bill are fully engaged in their work, how they feel about their jobs, and how they communicate and collaborate with others.

The "sense and purpose" has a huge power on the way we look at our tasks.[23] A Gallup report from 2017 provides employee engagement data from 155 countries around the world.[24] The results of Gallup's survey place employees into three categories defined by Gallup as follows:

- **Engaged:** Employees who are highly involved in and enthusiastic about their work and workplace. They are psychological "owners," drive performance and innovation, and move the organization forward.
- **Not engaged:** Employees who are psychologically unattached to their work and company. Because their engagement needs are not being fully met, they are putting time—but not energy and passion—into their work.
- **Actively disengaged:** Employees who are not just unhappy at work—they are resentful that their needs are not being met and are acting out their unhappiness. Every day, these workers potentially undermine what their engaged coworkers accomplish.

Table 3.1 displays selected results from the survey.

Table 3.1. Employee Engagement, Selected Regions

| **Region/Country** | **Engaged** (%) | **Disengaged** (%) | **Actively Disengaged** (%) |
|---|---|---|---|
| World | 15 | 67 | 18 |
| United States/Canada | 31 | 52 | 17 |
| Latin America | 27 | 59 | 14 |
| Southeast Asia | 19 | 70 | 11 |
| Australia/New Zealand | 14 | 71 | 15 |
| Middle East/North Africa | 14 | 64 | 22 |
| South Asia | 14 | 65 | 21 |
| Western Europe | 10 | 71 | 19 |
| East Asia | 6 | 74 | 20 |

The survey spanned across many different industry sectors and types of workers, and their degree of engagement is demonstrated in Table 3.2.

Table 3.2. Engagement by Type of Worker

| | Engaged % | Not Engaged % | Actively Disengaged % |
|---|---|---|---|
| Manager/Executive/Official in a business or the government | 28 | 63 | 9 |
| Professional: doctor, lawyer, engineer, teacher, nurse, etc. | 27 | 62 | 11 |
| Service worker: maid, taxi driver, maintenance or repair worker, etc. | 18 | 64 | 18 |
| Farmer/Fisherman/Other agricultural laborer | 18 | 60 | 22 |
| Clerical/Other office worker/Sales worker | 14 | 72 | 14 |
| Construction/Manufacturing/ Production worker | 12 | 64 | 24 |

The Gallup survey results show that a significant amount of human capital is unused and wasted both for organizations and for multitudes of individuals spending a considerable part of their lives doing things that generate little benefit to them as human beings besides a paycheck.

Issues related to people/workforce will be further investigated by the UAIM/WISE project. A team of UAIM utility partners will collaborate on several aspects of managing workforce:

- Conceptual models identifying issues related to people and describing how they impact employee engagement,
- Methods for assessment of current ("as is") employee engagement in an organization,
- Defining the characteristics of improved ("to be") workforce that would benefit both the individuals and the organization, and
- Developing guidelines for transition from the current to the desired state of workforce engagement.

This research will be led by the participating UAIM/WISE utility partners with support by the UAIM project team; it will draw on previous and current related research, methods, and tools within and outside of the water utility sector. The continued WISE research needs to achieve a more detailed and complete understanding of the "people" column, including issues related to workforce, organizational behaviors and culture, as well as change management. In Section 6.2 there is an outline of the research effort. People aspects are further discussed in Chapter 5.

## 3.3 Value Creation—Technology

Technology is seen primarily as an enabler of processes and decisions. A detailed examination of the technology component is not currently in the scope of the UAIM project. The WISE project is aiming for more in-depth considerations of technology. It is crucial to structure business processes and to educate and motivate people in such a way that technology can support the business processes to create value.

Except for WISE, the role of technology is studied in other programs. Several research efforts on technology aspects are ongoing under the umbrella of "Smart Utilities" including the WRF project 5039 "Smart Utilities and Intelligent Water System."

This aspect of technology is broken down into four layers representing the following capabilities:

1) *Sense and observe*. This capability could be provided by physical sensors (e.g., flow meters) and operator observations but also via questionnaires, surveys phone calls, or social media.
2) *Collect and structure*. After being received via the sensing layer, data need to be collected, checked and screened, organized, and structured. An example of this is asset registry.
3) *Insight*. Technology also offers analytical and statistical tools that can process data and provide insights—that is, produce useful information from loads of data.
4) *Decision support*. Data and timely and reliable information can be used to improve decision making.

Note that this structure relates directly to the feedback principle discussed in Chapter 2 (Figure 2.1 and the box on *feedback*, Chapter 1). We can use different assessment methods and maturity models for each of these layers or capabilities. For example, we may have deployed sophisticated instruments for a certain process, but the data received from these instruments is not properly organized, or technology may provide effective decision support on the local level (e.g., maintenance unit), but the information is not available for business processes outside that specific system or business unit.

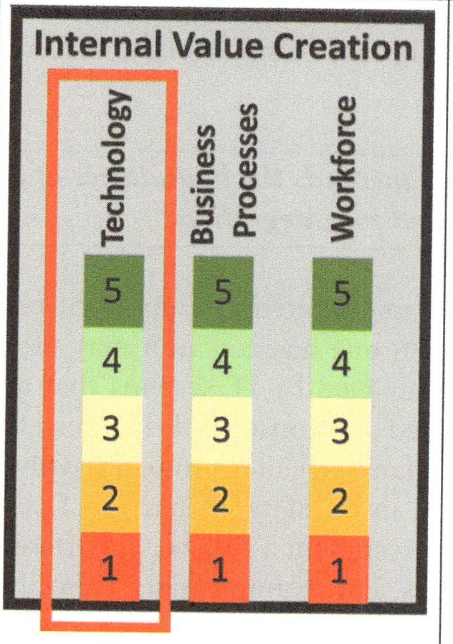


From Figure 3.1

A key added value that can be obtained by using technology is improvement in the quality of our decisions. Improvements in decision making can be accomplished in two different ways, *automation* and *decision support* (see Section 2.2).

We remind that automation is more prevalent on the operational level while decision support is more common on tactical and strategic levels where humans still have the primary role in making decisions.

***It is crucial to structure business processes and to motivate and educate people so that technology can support the business processes to create value.***

A typical "landscape" of technology solutions within a water sector utility includes a number of technologies such as:

- computerized maintenance management systems (CMMS),
- laboratory information management systems (LIMS),
- GIS,
- computer aided design and drafting (CADD) systems,
- real-time control systems such as SCADA and distributed control systems (DCS),
- automatic meter reading (AMR) or automatic meter information (AMI) systems,
- project management systems,
- field instrumentation, and
- back-office applications such as financial systems, payroll, and human resources systems.

Many functions, business processes, and business units within a utility are critically dependent on properly working technology systems. A number of these technologies offer their primary users with extensive capabilities to address their own business challenges within a specific area.

The key hurdles are those of integration: they become apparent when a utility is facing an "enterprise-wide" challenge that transcends the boundaries of a single organizational unit.

| |
|---|
| ***A utility facing an "enterprise-wide" challenge that transcends the boundaries of a single organizational unit will face the need for integration.*** |

Addressing issues related to lack of technology integration is often an important part of improving business processes. Decision making within one business unit may often be improved if the information that is collected and managed by IT systems that are "owned" by other business units and can be easily accessed and considered. The need for technology integration is especially apparent for asset management because it involves many business units, including engineering, operations, maintenance, finance, IT, and CIP planning; data related to asset management systems resides in a number of different IT systems and is often not available to secondary users. For example, data regarding maintenance may not be readily available to operators or planners in engineering.

An important consideration is the need to address the rational for change. In the groundbreaking book on organizational change John Kotter[25] noted, "It must be considered that there is nothing more difficult to carry out, nor more doubtful of success, nor more dangerous to handle, than to initiate a new order of things." The lesson learned from past efforts in adopting and implementing the technologies is that advancements in technology in and of themselves are not sufficient reasons to change. Many technology initiatives have failed because they fundamentally failed to address the need to change and the organizational readiness to adapt people and process.

| |
|---|
| ***Many technology initiatives have failed because they fundamentally failed to address the need to change and the organizational readiness to adapt people and process.*** |

An online survey in 2011 of 45 wastewater utilities, conducted by Cello Vitasovic,[26] included questions about their use of technology, in particular IT systems. The responses provided several conclusions:

- IT systems are providing support to many business areas and have become an indispensable component of managing/running a wastewater utility.
- Some of the IT systems play critical roles in the business of a utility, and their failure would cause serious disruption of essential functions in a utility.
- Most of the IT systems are providing value to utilities, but not to the same degree. Systems that are directly involved in real-time operations (e.g., SCADA) or maintenance (CMMS) are seen as providing the most value.

- Different types of problems are common on projects dealing with IT. Projects dealing with real-time control systems (e.g., SCADA, DCS) experience the most problems.
- Some IT systems constitute an important part of the infrastructure that the organizations rely upon. The importance of interpersonal communication, collaboration, and information sharing, is underscored by the fact that email/calendar systems are reported to impact most business functions, including operations and maintenance.
- Data sharing and integration among IT applications is lacking.
- As emphasized in Chapter 1, there is a well-recognized difficulty to integrate different key software solutions. The trend is a further fractionating and budding—leading to increasing complexity and need for mastering many different systems.

Reliability of digital tools ought to be high on the agenda in future studies. In particular, cybersecurity should be prioritized. Implementing effective cybersecurity will be increasingly challenging because devices, people, and attackers are becoming more innovative. Plans for future studies are summarized in Section 6.2.

A more detailed outline of technology in WISE is presented in Section 6.3.

## 3.4 Integration—The Core of Systems Thinking

Integration is a key challenge. This becomes apparent when a utility is facing an "enterprise-wide" task that transcends the boundaries of a single organizational unit. Lack of internal integration is often an obstacle to manage complex problems. Here we consider two aspects of integration:

- integrating people and processes, and
- integrating technology into the organization.

***Lack of internal integration is often an obstacle to manage complex problems.***

Integration is the core of systems thinking. The same underlying idea of "systems" can denote technical, social, economic, and living systems. The very meaning of systems derives from the Greek *syn* + *histanai*, meaning "to place together." To understand things systemically literally means to put them together to establish the nature of their relationships. Complex organizations, technical processes, or economical activities can be described by their smallest components "plus organizing relations."

A utility may be described by a form of multileveled structure of systems within systems. Each of these subsystems can form a whole with its parts, while they are also part of a larger whole. So we have different levels of complexity with different kinds of relations at each level. The notion of microorganism activity or flow rate is important at the technical process operating level but has limited value for strategic decisions at the management level.

During a UAIM project workshop held in September of 2019, the roughly three dozen representatives from different utilities discussed four topics that they had selected for collaborative effort on business process modeling, analysis, and improvement:

- delivering CIP,
- developing asset management plans,
- managing enterprise risk, and
- business case evaluation and prioritization (for the CIP).

As part of the workshop, four groups of participants identified key challenges related to each of these topics. After the challenges were identified, each of the challenges was placed on the 3×3 UAIM Framework matrix shown in Figure 1.2: that is, if a challenge was deemed to be mostly related to technology on the operational (real-time) level, it would be placed in the lower right corner of the matrix. The overall results across these topics are shown in Figure 3.3.

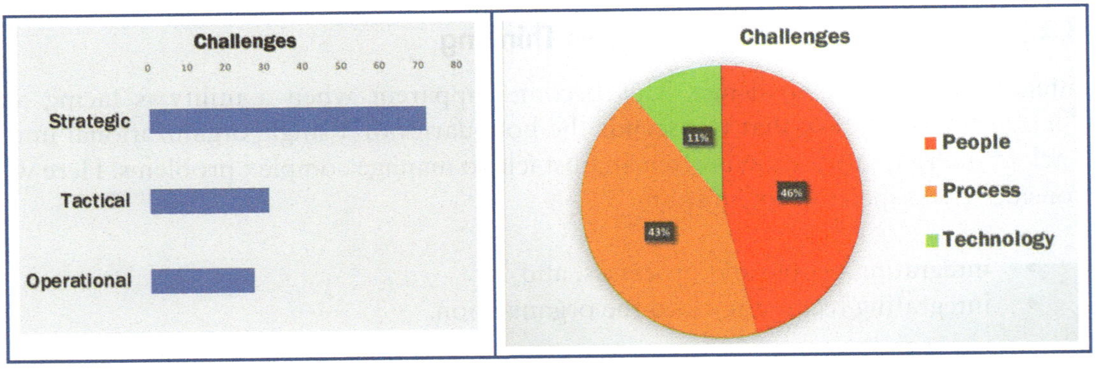


Figure 3.3. Challenges in Key Business Functions in Water Sector Utilities

The results in Figure 3.3 illustrate that the root of the challenges in addressing key functions in water sector utilities are seldom related to technology itself but occur when technology is introduced into a broader context of an organization that includes people and processes. Gerald Kane from Boston University and a team from Deloitte[27] surveyed many companies in the private sector to examine how they are leveraging modern technologies to gain competitive advantage and concluded "people are the real key to digital transformation."

***Key challenges in water sector utilities are seldom related to technology itself but occur when technology is introduced into a broader context of an organization that includes people and processes.***

The findings of the Kane group have been further studied particularly for the water sector by Vitasovic and Karimova.[28] Their results have led to the inclusion of workforce, organizational culture, and change management as a major part of UAIM scope for its final (fourth) year as a WRF project.

In Sections 5.3 and 5.4 we will further discuss what forms the basis for decisions by individuals and by organizations.

# 4 External Impact

| Key Messages | |
|---|---|
| | • Availability and effectiveness of water infrastructure influences every person's daily life.<br>• Utilities are traditionally *production* oriented, but there is a development toward customer- or *demand*-oriented operations.<br>• With increasing water scarcity, the issue of *water reuse* becomes increasingly important. Required water quality related to water use will be on the agenda.<br>• It is a challenge for any utility to influence *water consumption* as a way to avoid or delay major investments in production capacity.<br>• *Tariffs* or *rates* ought to reflect both the production costs and the true economic value of water services. How to combine the goal that water is a human right with the requirement to operate a water utility economically?<br>• The Maslow hierarchy model demonstrates that water is fundamental to every person, but the model can also characterize a water utility organization. |

A water utility must measure the overall value that it delivers to the outside world: that is, the quality (maturity) of impact that it has on external entities to ensure that informed decisions are made based on data, knowledge, and wisdom, as depicted in Figure 4.1. Water scarcity is becoming a reality in many regions and will force utilities to attend to and influence customer behavior, making the operation customer oriented, Section 4.1. In Section 4.2, we consider various possibilities to secure future water supply, including water reuse, customer behavior, and water tariffs. The value of water can be related to human needs, as illustrated by the Maslow hierarchical model, Section 4.3. Access to safe water is so much more than satisfying the basic physiological needs. It will improve health and is fundamental for food production as well as energy production. Section 4.4 further discusses how water operations have to adapt to both people and environment expectations.

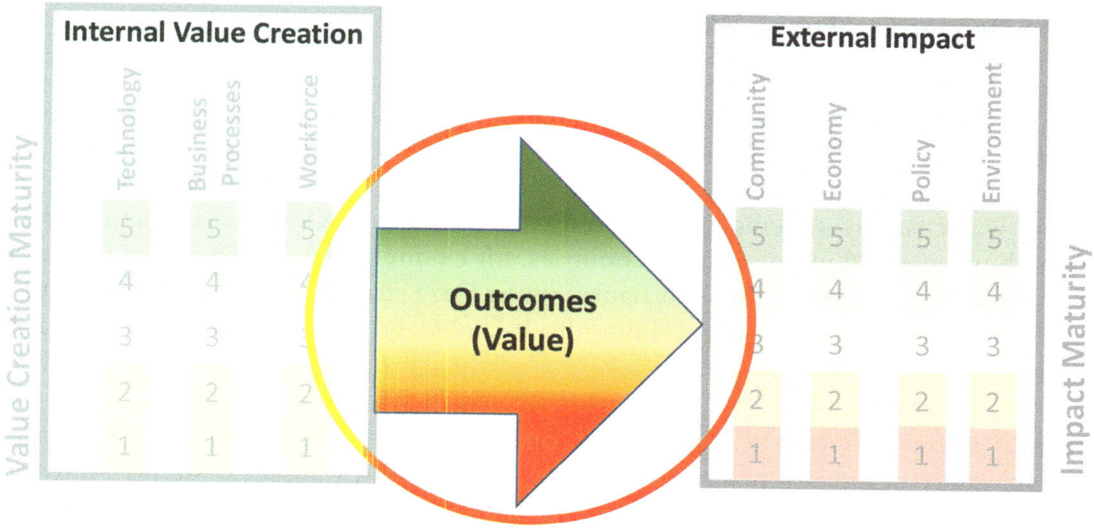


Figure 4.1. Value Created by a Water Sector Utility

## 4.1 The Outcome

Since a water utility usually has a regional monopoly, the value of the company must be defined somewhat untraditionally. Water scarcity is becoming a reality in more and more regions and the issue of sustainable operations will be increasingly important. It is estimated that two out of every three people will live in water-stressed areas by the year 2025. Clean water supplies and sanitation remain major problems in many parts of the world, with 20% of the global population lacking access to safe drinking water. Water scarcity is not limited to low-income countries but is a reality in an increasing number of regions in high-income countries. The value of water will increase, and there will be a challenge to match the value with tariff structures.

The traditional goal of water supply operations has been *production* oriented. This implies that the utility is all the time trying to satisfy the customer demand. Water availability must be ensured for a long time, causing sustainability and resilience to be vital. It also means that the water demand has to be satisfied for coming generations. However, it is no longer apparent that enough water can be supplied to satisfy any demand of water. Therefore, the utility should increasingly deal with customer demand, behavior, and attitudes. The *production*-oriented utility ought to be gradually changing toward *customer* or *demand* oriented. Trust becomes a key issue.

***Any utility should increasingly deal with customer demand, behavior, and attitudes. Trust is a key issue.***

## 4.2 Safeguarding Future Water Supply

A water-wise utility recognizes that it depends on nature, as emphasized in Section 2.2. Do we pay the right price to extract water, and do we pay nature what it costs to return the treated wastewater to the receiving water body? Water supply sometimes is obtained from aquifers having fossil water that is not renewed within generations. Too many water supply systems are built around non-sustainable water extraction, both in high-income and in low-income countries, for example in northern China (Beijing), India (Gujarat, Chennai), the West Bank and the Gaza strip, South Africa (Cape Town), and southwestern United States. Logically it should be more expensive to extract a nonrenewable water source than a renewable one.

If local water resources are insufficient, one alternative is to pump the water from distant sources, already practiced in many regions, for example to supply water to Southern California, South Australia (Adelaide), South Africa (Johannesburg), or China (Kunming in the Yunnan province). Huge water transfers from the Yangtse River toward northern China is an example of massive water transportation. Not only the cost of transporting the water, but the environmental cost is substantial, both in terms of building the pipelines and the energy use for the operation. If there would be a carbon emission tax, this cost would be significant and would influence the water tariffs.

As an alternative to long water transports, seawater desalination is applied in many areas. The energy issue is usually brought up as the obstacle for desalination. However, with energy requirement of 4–5 $kWh/m^3$ for desalination, the energy cost is not significantly higher than the water transfer cost of 1–3 $kWh/m^3$ in some regions. Furthermore, if renewable energy—solar PV or wind—can be used, then the environmental price for desalination would be lower than long water transfers.[29] Today, the global desalination production corresponds to around 12 L/d per person of the global population of 7.8 billion people.[30]

### 4.2.1 Water Reuse

Traditional water supply provides drinking water quality for all kinds of water use, from food preparation to toilet flushing. With increasing water scarcity, the issue of water reuse becomes a necessity in many places, and the concept of required water quality related to water use is considered in many places.

> ***With increasing water scarcity, the issue of water reuse becomes a necessity in many places.***

Treated wastewater is an effective alternative water supply. According to a National Academy of Science report,[31] expanding water reuse—the use of treated wastewater for beneficial purposes including irrigation, industrial uses, and drinking water augmentation—could significantly increase the available water resources.[32] Reusing water after appropriate treatment extends its life cycle, thereby preserving water resources. The acceptance of water reuse is deeply coupled to public trust, as further discussed in Section 4.3.

### 4.2.2 Influencing Customer Behavior

A goal for a demand-oriented utility should be to influence water consumption so that some major investments in production capacity can be avoided or delayed. Can customers be motivated to limit their water use? This is of course quite a different business goal than for a traditional private enterprise. There are different ways to influence the demand:

- water saving equipment,
- water metering,
- water tariffs,
- customer attitudes, and
- water reuse.

For a long time, utilities have encouraged or subsidized water saving equipment like showerheads, faucets, toilets, washing machines, and dishwashers. With increasing water scarcity incentives for using metering, often with advanced metering technology, will increase. So-called smart water meters, connected to a wireless network, have made the water use decrease by between 2.5% and 29% in various locations.[33] Thames Water supplies water to 3.3 million properties in the London area. The utility has an ambitious smart metering installation program and aims to have metered 100% of connections across the region by 2030. Presently around one third of the customers have meters installed, and they use around 12% less on average compared with the customers without individual metering. Smart measurements at home will monitor normal as well as abnormal consumption and place customers in a position to manage their use to avoid peak charges or to adapt to tariff pricing structures.

> ***How can a demand-oriented utility influence water consumption so that some major investments in production capacity can be avoided or delayed?***

A key challenge is to build trust between the utility and the customers. One of the reasons for high leakage in many urban areas, particularly in low-income countries, is that people are paying too little for the water. If a shower or a toilet is broken, the water is allowed to leak. Efficiency is primarily obtained by systematic and simple maintenance, repair, and operation.[34]

### 4.2.3 Tariffs and the Value of Water

Access to clean water is recognized as a human right. This immediately leads to the question: how "much" is a human right? It is not a human right to waste water. Clean water requires treatment, and a vast infrastructure and distribution do not come cheap. In many countries, it is undertaken as a public utility fully paid for by the taxpayer. However, the cost of accommodating growing demand is too high for the public purse. The affordability of water has become an urgent issue in many places, also in high-income countries.[35]

If water is priced correctly according to market evaluations, there will be some who will have to use water so frugally it could pose a danger to personal and public health. In poor countries, even in the middle of cities, people are resorting to untreated groundwater for the daily needs. In these circumstances, outbreaks of cholera and other waterborne diseases are a regular occurrence, especially where a proper sewerage system is lacking. For public health and other reasons, every household should have access to treated, piped water—but at a price structure that incentivizes conservation.

In many countries, both low-income and high-income regions, water is massively subsidized, subjecting it to serious under-valuing and severe misuse by individuals and industry. Yet, there is a need to ensure that every human will have the right to get clean water. Water pricing needs to be revised in many places and countries. The closer the price of water approaches full cost of service, the better water can be valued.

Many places charge the water so that even the poorest people can afford a minimum amount of water use, the most valuable water that life depends on. Affordability is a major issue, and many cities and regions have resorted to utility credit programs or other means in their tariff structure to achieve affordability goals.[36] Anything above this level should be priced according to the real costs it takes to make the water drinkable. To water a lawn in a water-scarce area is not a human right and should be charged accordingly. Simply expressed, we should pay less for the necessary water need and more for the "luxury" needs. This is qualitatively illustrated in Figure 4.2.

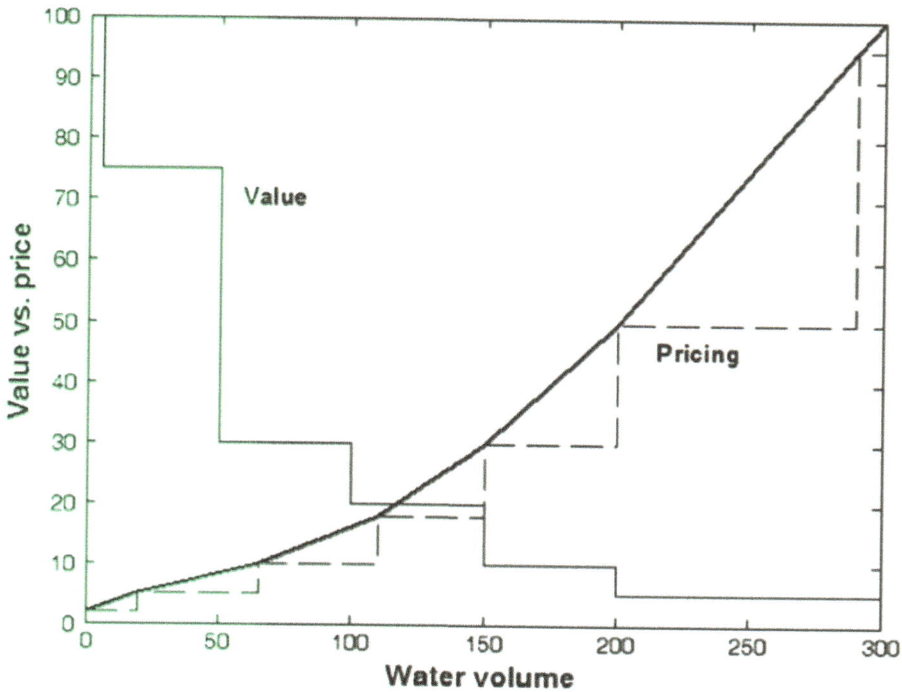


Figure 4.2. The Perceived Value of Water Compared to a Structure of the Tariffs[37]

It is crucial to find some technique to determine the economic value of water services. Of course, this is related to the willingness to pay. It looks as if too many countries and governments consider water as a limitless natural resource that can be freely exploited and used by any authority or by the landowner. Opposite to any other commodity—such as oil—there is hardly any market defined for water. Only the cost of pumping, treating, and distributing the water is commonly charged. There is mostly no cost specified for the degradation of the water ecosystem. To pump the water from a river or other surface water makes no difference than pumping fossil water from an aquifer. The water seems to belong to everyone, and nobody has the responsibility.

*Mostly, there is no cost specified for the degradation of the water ecosystem. Only the cost of pumping, treating, and distributing the water is commonly charged.*

Some countries have recognized that the tariff should encourage efficient use of the water and have a tariff structure motivated by Figure 4.2. This is the practice in Greece, in China, in India, and in many African countries. Figure 4.3 shows progressive tariffs in some Asian and African cities. The tariffs are continuously updated, but the principle remains.

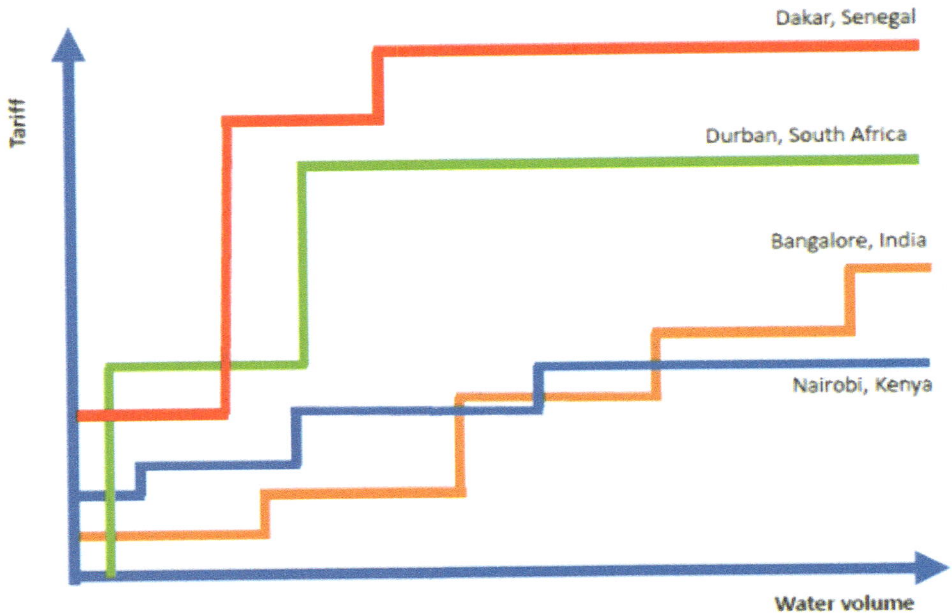


Figure 4.3. Progressive Tariffs in Some Cities in Africa and Asia[38]

The other extreme structure of water tariffs is applied in many places, where the fixed connection cost may be as high as 90% of the total water cost. Naturally, this tariff structure does not provide economic incentive for the user to save water.

***How to combine the goal that water is a human right with the requirement to operate a water utility economically? Tariff structure should consider this.***

Water tariffs should reflect not only production costs but also water and energy availability. Conventional economy does not calculate the cost for water degradation. The economy of water sometimes resembles the economy of ecology. Biodiversity brings stability to ecosystems, which provide a wide range of "services" that businesses rely on yet received free of charge. Because there is no financial cost for these services, they have been treated as being without value. This has resulted in corporate decisions that damage the ecosystem, reduce biodiversity, and leaves fewer degrees of freedom for future action. The solution is to value these ecosystem services so that they can become part of planning and decision making. This has nothing to do with corporate social responsibility or the green agenda, it is hard-nosed economics. If the economic values of those services are considered, decisions automatically promote sustainability.

## 4.3 External Impact

As UAIM transitions to WISE, future efforts will include examination of different metrics and key performance indicators for water sector utilities. Increasingly, global sustainability metrics are being used, and sustainability measures are now being put in place and advocated by leading industry organizations like American Water Works Association (AWWA) and WEF.

The availability and effectiveness of water infrastructure impacts every person's daily life. Using Maslow's Hierarchy of Needs[39] and his view on human motivation helps framing how these impacts are felt. Maslow's hierarchy posits that people care about their basic physiological needs first (the base of the pyramid in Figure 4.4). Once these basic needs are satisfied, they move up the hierarchy to those aspects of their world that impacts their safety and security. After those needs are satisfied, people then seek out the higher levels of love/belonging, esteem, and self-actualization.

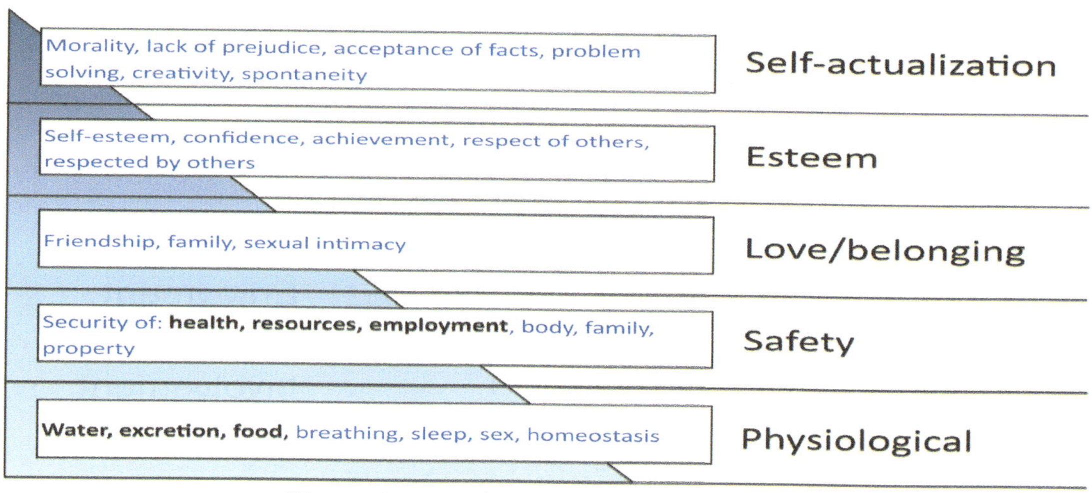


Figure 4.4. Maslow's Hierarchy of Needs

Water is explicitly identified in the physiological level, with water (drinking water supply), excretion (sanitation services and wastewater treatment), and food (irrigation). The success of water and wastewater infrastructure to reduce waterborne illness improves health on the second step of the hierarchy. Without sufficient water availability, resources anywhere from agriculture to energy production would be limited.

If Maslow's hierarchy demonstrates water is fundamental to every person, we can also apply a similar model to an organization—the water utility. Public health and environmental health are the foundation of the utility's core mission, analogous to the bottom level of Maslow's pyramid. UAIM's definition of external impact of the water utility also addresses higher levels of needs, moving from the environment (public and environmental health) to policy, the economy, and overall community.

> ***Maslow's hierarchy describes how water is fundamental to every person. Similarly, public and environmental health are the foundation of the utility's core mission.***

While water is prevalent in the base of the hierarchy, the higher levels of needs are generally not recognized as directly related to water. However, if the framework is applied to IWRM or total water management, water and wastewater utilities can develop a set of metrics that incorporate the higher level needs shown in Figure 4.5. Keeping the bottom two levels relatively the same, access to water/sanitation is equivalent to Maslow's physiological level. Reliability, resilience, and safety in the IWRM adaptation parallels the safety and security levels. Public participation and involvement in water issues maps to the belonging level of Maslow. Recreation and other beneficial use of water resources and environmental justice represent how the esteem and self-actualization levels in Maslow can be applied to IWRM.

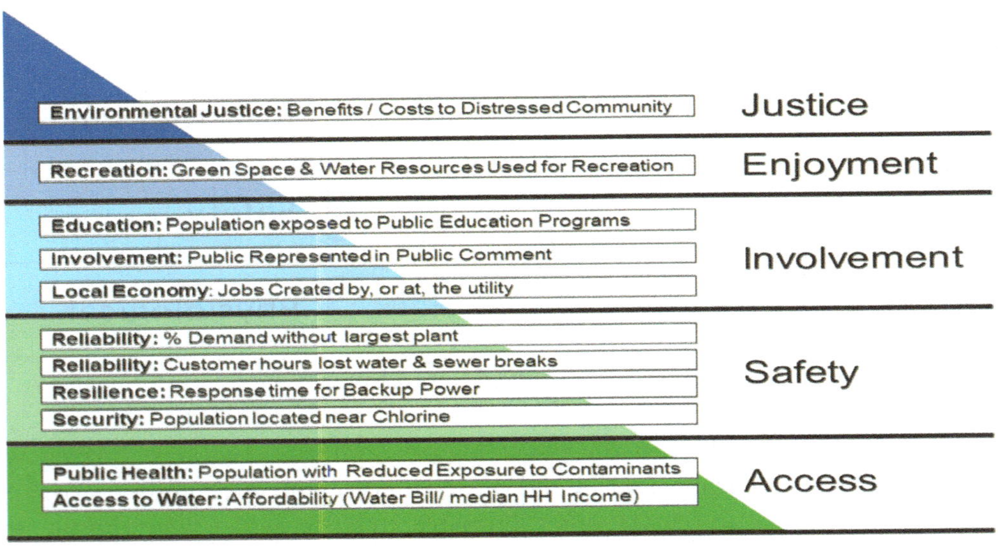


Figure 4.5. Adaptation of Maslow's Hierarchy to Integrated Water Resources Management (Liner et al., 2012)

## 4.4 Adaptation to Human Society and Environment

A critical component of both the IWRM adaptation is the focus on the involvement layer in the middle. From public education to community events like tree planting, public participation is critical for the success of utility stakeholder engagement. This authentic engagement with the public uses the concepts of "legitimacy," which has been applied successfully for the development of water reuse projects, particularly those involving potable reuse (see Section 4.2.1).

Orange County, California, may exemplify how legitimacy can produce successful results. Orange County Water District (OCWD) and Orange County Sanitation District (OCSD) jointly developed a Groundwater Replenishment System (GWRS).[40] The utilities put an extensive plan in place to authentically engage the community, not simply a marketing effort. During ten years of development, the utilities hosted community engagements including tours for the general public, public officials, regulators, and environmental groups. The GWRS began operation in 2008 and currently produces around 100 million gallons per day (380,000 $m^3$/day) of high purity water. The final expansion will increase production by another 30%, enough to meet the needs of approximately one million people in their service area.

A study about the success of the GWRS in Orange County showed that the success of their indirect potable reuse program was due to the legitimacy of their effort. Through its dedication to the outreach efforts, utility managers were recognized as trustworthy and competent experts in the community. Truly demonstrating the impact of legitimate public engagement, the project served as the impetus behind having the California legislature pass a law allowing high purity water to be bottled for educational purposes beginning in 2017.

Orange County successfully demonstrated the three levels of legitimacy that need to be addressed to have a successful project. The basic (pragmatic) level focuses on the user's self-interest, seeking to answer questions such as "How do I benefit personally?" and "How am I involved in making the decision?" The next ("moral") level deals with social values and welfare, addressing questions like "How are safety and quality guaranteed?" and "Can I trust the organization?" The upper ("cognitive") level deals with customs and routines that are taken for granted, answering questions like "Does the technology fit with my normal daily life?" and "Is the technology essential, that is, with no feasible alternatives?"

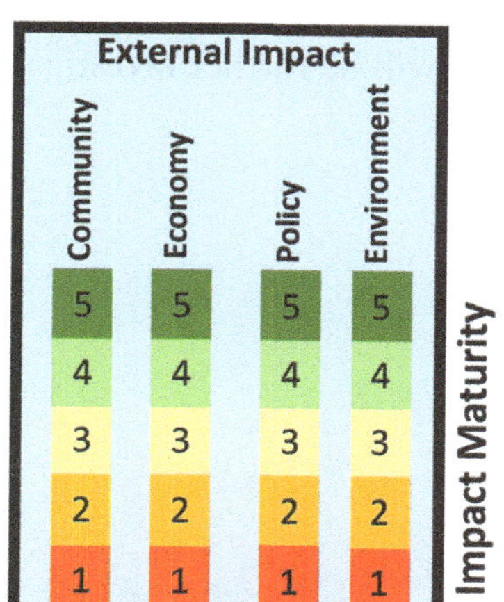


**From Figure 4.1**

The utility should adapt to the environment and the human society in a sustainable fashion. We ought to try to understand our human infrastructure systems as an integral part of the nature we inhabit because we live in a never-ending symbiosis with the environment. We accept that water is a human right, but we should also recognize that water is our responsibility. In our high-income countries, we too often take water accessibility for granted and forget that our lives depend on it. We must make sure that our environment, our systems, our organizations, and our usage are working in synchronicity. This is the challenge of this century. This is not in conflict with the original aim of "smart" or applying instrumentation, control, and automation to make operations more efficient and less expensive. It is a much loftier goal to reach true sustainability, and the task is much more complex, requiring a deeper understanding of our infrastructure systems; the natural, physical, and biological systems they interact with; and our usage. We need water wisdom[41] to understand where and when we must be careful, when to save water, and when to be ready with emergency plans.

***We accept that water is a human right.***
***We should also recognize that water is our responsibility.***

Wisdom and knowledge should guide us as we try to answer many important questions related to the external impact:

- How can we combine the goal that water is a human right with the requirement to run the operation of a water utility economically?
- How can we make the users environmentally aware as well as how can the utility show environmental awareness?
- How can we better understand and address social justice and affordability of clean water?

Maslow's hierarchy can be a tool to help define the maturity of the various aspects of external impact. These concepts will be further investigated in the ongoing research, as outlined in Section 6.2.

# 5 Organizational Learning and Change Management

| | |
|---|---|
| **Key Messages** | • Feedback plays a crucial role in utility operation operating, in particular to assess and learn from the impact on external users and nature.<br>• A key characteristic of a wise utility is its ability to learn and improve. |

We discussed qualities of a wise utility in Section 2.2, stressing the importance of feedback in operating a utility. This is further emphasized when the utility will assess its impact on the external users and nature, as depicted in Figure 2.2. In this chapter, we will further explore how the utility can learn from the external information to increase its value. We distinguish between single-loop feedback in Section 5.1 and double-loop feedback in 5.2. The latter is crucial for the organizational learning process. Building a wise organization requires that we understand how humans make decisions. This is explained in 5.3 and will lead to the discussion in Section 5.4 how organizations make decisions. Section 5.5 further discusses the road toward a wise utility. A major challenge is how to deal with "wicked" problems that involve many people and opinions, or economic obstacles, Section 5.6. To handle such problems requires wisdom, since there is no apparent correct solution.

## 5.1 Addressing the Challenge of External Adaptation

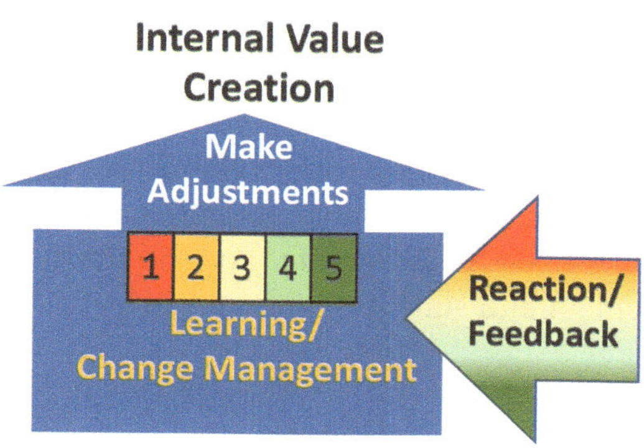


From Figure 5.1

An organization generates outcomes that provide value to external entities. External entities react to the value they are receiving from the organization and provide feedback that reflects the difference of the value they received and the value they expected or desired (compare Figure 2.2). The reaction from external parties provides valuable information to the organization and is a key component of tracking organization's overall performance.

For a private for-profit organization, such feedback may be easier to measure and track—for example, by the market response and performance against competition. As remarked in Section 3.1, the mission of a public water sector utility is multifaceted, and the key performance measurements are not focused on the market and competition.

The reaction from external entities may therefore also be more complex for a public utility as it may include a number of nonmarket factors such as affordability, social justice, satisfaction with utility performance, effectiveness in implementing capital and service delivery, engagement opportunities, and stewardship of precious natural resources (Section 4.2.3). It is important to note that different aspects of impact also have different constituents.

A key characteristic of a wise utility is its ability to learn and improve. Organizational learning includes the capability of an organization to:

- *track its performance* based on the reaction from external entities—including all the components of reaction from the external impact, and
- *manage changes* to its internal value creation (people, processes, and technologies) and their impacts, required to adapt its programs, priorities, and behaviors based on external reactions (feedback).

Tracking performance includes monitoring different reactions from external entities to learn the impact that the organization is having on the outside world.

A common type of learning is represented by a single-loop model of learning that includes process of adjusting our strategies and techniques based on the impact of our actions. This single-loop learning applied to the WISE model is illustrated in Figure 5.1 (see also Figure 2.2). The feedback information from the impact and result going back to strategies and techniques contains information about the difference between the expected and the actual outcomes.

As discussed in Chapter 2, feedback is a powerful concept.[42] In an automatic control system of technical processes, the measurements are provided with online sensors. Here the "measurement" is monitoring and evaluation of the evaluated outcomes of the utility. In the technical process, the control decision is automatic. For the utility, the decision is made via learning on the management and staff level.

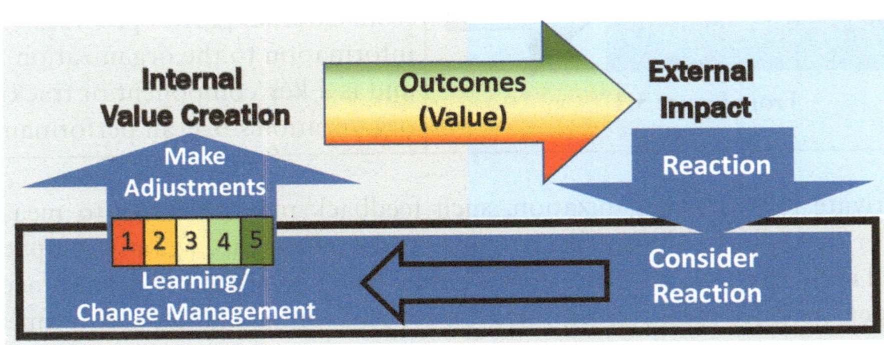


Figure 5.1. Single-Loop Learning Within the WISE Framework (Detail of Figure 2.2)

Single-loop learning is common; this reactive model works well when applied to simple "loops." However, it may not work as well for managing interactions between more complex systems. The scope of the UAIM/WISE project includes examination of change management in water sector utilities, as described in Section 6.2.

## 5.2 Organizational Culture

Organizational culture is a critical and overly complex aspect that drives an organization's behaviors. This section discusses the impact of organizational culture on organizational learning and change management.

Figure 2.3 with details copied in Figure 5.2 illustrate the concept of double-loop learning, represented within the WISE Framework. The single-loop learning (Figure 5.1) will adjust strategies and techniques, while the double-loop learning brings in a feedback signal to our basic assumptions that include our underlying values and beliefs that drive our behaviors. Each of the variables is dynamic—it changes in time.

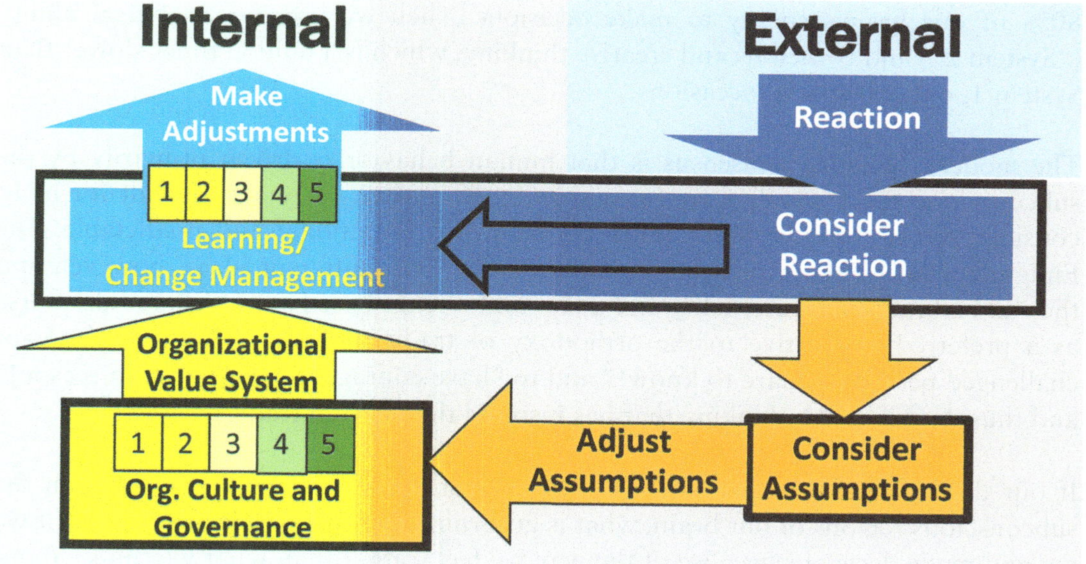


Figure 5.2. Double-Loop Learning Model Within the WISE Framework (Detail of Figure 2.3)

| |
|---|
| *Responding to external reactions includes not only adjusting our strategies and techniques but should also consider the underlying values and beliefs that drive our behaviors.* |

The scope of the UAIM/WISE project will include examination of organizational culture in water sector utilities. This research will include a maturity assessment method, a survey, and a review of the existing research into organizational culture both within and outside of the water sector, Section 6.2.

## 5.3 Understanding How Humans Make Decisions

For an individual, a critically important question is about their existence in the world, a question about purpose: "Why am I here?" Throughout history, many brilliant people have provided us with clues. In the first part of the twentieth century, three psychologists who lived and worked in Vienna provided three different answers: Sigmund Freud thought that man seeks pleasure, Alfred Adler though that he seeks power, and Viktor Frankl[43] thought that man searches for meaning. The origin of the word "individual" comes from "indivisible," and it should be understood that every person is different from each other, and no single theory could possibly describe all human beings.

Recent advances in psychology and neuroscience are teaching us that our decision making is governed by the older part of the brain that rules our subconscious behavior.[44] We follow our "gut," called System 1, which is fast and intuitive thinking that consumes 80% of our brain's energy to make decisions. Then we use our analytical ability ("System 2") and conscious and creative thinking, which is a million times slower than System 1, only on special occasions.

The modern scientific consensus is that human behavior is driven primarily by the subconscious processes that are beyond rational control. They are difficult for us to consider because we prefer to think of the world as rational and predictable. The Enlightenment thinkers such as Immanuel Kant[45] had different ideas, however, and they all embraced the rationality and a fact-and-science based approach and discourse as a preferred alternative to the orthodoxy of traditional established beliefs. Kant challenged people to "dare to know!" and to "have courage to use your own reason!" and thus described the thinking that has inspired the Western tradition.

If our decisions and behaviors are driven by instinctive systems that operate in the subconscious regions of our brain, what is governing our subconscious instincts? If we act and make decisions because of the way we feel (rather than what we think) about them, what are the drivers that determine how we *feel* about different things?

A system of values and a sense of purpose are of great importance for a human being. A person's moral foundation will define how he or she perceives issues and acts in different situations.

***A system of values and a sense of purpose are of great importance for a human being.***

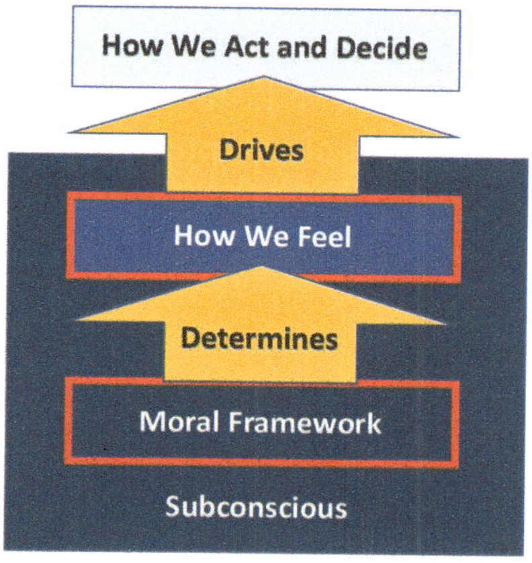


Jonathan Haidt[46] suggested that the foundation that governs our subconscious feelings is our moral framework: it defines how we feel about a wide range of concepts and issues. Haidt initially identified five components (pillars) of the moral framework and later added the sixth one:

1. Care/Harm
2. Fairness/Cheating
3. Loyalty/Betrayal
4. Authority/Subversion
5. Sanctity/Degradation
6. Liberty/Oppression

Critics have pointed out that the list is incomplete and that some items should be added. For our purpose, the accuracy or completeness of the Moral Foundation Theory is not the main point: the key point is the critical importance of a person's moral framework—system of values. The six pillars of the moral framework are not equally strong: people attribute different levels of importance to each of these pillars. Through surveys, Haidt showed that the pattern of importance (i.e., how strongly a person feels about each of these pillars) indicates a person's system of values and predicts their political leanings. Liberals, for example, place much more importance to the Care/Harm components while Conservatives place much more importance on Sanctity/Degradation than Liberals. For Libertarians, the sixth pillar is exceptionally strong.

Decision science considers decision making from two different perspectives: *prescriptive* (how to make best decisions) and *descriptive* (how human beings actually make decisions). The prescriptive perspective follows the rational model (using the logic of consequence): it examines the consequence of each decision and ranks the choices based on their impact according to some measurable criteria. The rational decision model includes the following steps:[47]

1. identify and define the problem and the goals,
2. generate list of alternatives (solutions),
3. define the assessment criteria (for evaluating the solutions),
4. evaluate the alternatives and choose the best one,
5. implement the preferred alternative,
6. monitor and evaluate outcomes and results, and
7. feedback—make adjustments to future actions based on outcomes.

This method can be successfully applied to certain tasks. Examples include event-based processes such as playing a chess game, where every move is judged by its future consequences. A very simple example of such reasoning in the continuous time domain is control of temperature in a room—for example, a thermostat measures the temperature in the room (the consequence of many factors and a system state indicator that can be easily measured) and instructs the heater to turn on or off. Several equipment or plant unit control systems are rational, like level control, pressure control, flow control, and dissolved oxygen control.

The rational model, however, has been notoriously lacking when applied to complex problems such as economics when the behavior of (actual) people played a big part in the process. Economy is merely one aspect of the whole ecological and social fabric. Economists tend to neglect this social and ecological interdependence, treating all goods equally without considering the many ways in which these goods are related to the rest of the world.

***A rational model has been notoriously lacking when applied to complex problems such as economics.***

For a certain category of problems, especially those that include human interactions, a descriptive view of decision science may be more appropriate. According to this model, humans do not follow the rational model, and an individual will make decisions primarily based on the responses to the following three questions:

1. Who am I?
2. What kind of situation is this?
3. What is a person like me supposed to do in a situation like this?

Quite often individuals involved in the decision process place different values on the decision criteria, and frankly operate out of fear or lack of trust or are just thinking about personal impacts on their role and positional authority or have biases that may or may not be shared with team members. Others may feel afraid to speak up or feel threatened, or power is not appropriately allocated equitably.

According to this (descriptive) perspective, human behavior follows the model of identity because the answer to the question "Who am I?" is the foundation for our decisions. During our lifetime, we develop and acquire different identities; with each new identity, we learn different patterns of rules and situations when we make decisions. Developing an identity is the fundamental part of the educational process. If we join the army, they will give us a haircut and a uniform, assign the identity of a soldier, teach us about the situations a soldier may encounter, and instruct us in how we should behave is such situations. We undergo the same process if we are acquiring an identity of a medical doctor, mother, water treatment operator, water customer, water utility manager, regulator, or an environmental activist.

Our values, morality, and identity are shaped through our interactions with different individuals and groups that include our family, friends, coworkers, bosses, and many others. Through these interactions we learn the unspoken/unwritten rules of behavior that fall within an overarching concept of our culture. Human beings' actions and reactions on issues that really matter to us are based on feelings rather than on rational thought, and we are guided mostly by pattern recognition rather than rational analysis. We need to learn about the personal sense of identity, values, and culture if we want to successfully predict or influence the behavior of individuals.

***Our actions and reactions on issues that really matter to us are based on feelings rather than on rational thought.***

Management of a water sector utility includes making many decisions on strategic, tactical, and operational levels. As shown in Figure 3.3, 46% of the internal challenges were considered related to "people" aspects (i.e., workforce and organizational culture). Wise (knowledge-based) decisions related to those aspects require us to develop a good understanding of the culture and workforce issues and leverage that understanding to improve decision making.

## 5.4 Understanding How Organizations Make Decisions—Organizational Culture

Already in Section 2.4 (see also note 25) we paid attention to organizational culture. A sense of value is a critical component of an organization's culture. Table 5.1 describes three levels of culture.

Table 5.1. Levels of Culture (Schein & Schein, 2017)

| | | |
|---|---|---|
| Artifacts | Visible and feelable structures and processes | Difficult to decipher |
| | Observed behavior | |
| Espoused beliefs and values | Ideals, goals, values, aspirations | May or may not be congruent with behavior or other artifacts |
| | Ideologies | |
| | Rationalizations | |
| Basic underlying assumptions | Unconscious, taken-for-granted beliefs and values | Determine behavior, perception, thought, and feeling |

Organization's espoused sense of purpose may be stated in "mission statements" and strategic plans. The true values of an organization and the primary drivers of behavior are contained in the bottom layer (basic underlying assumptions) of organization's culture. Just as an individual's values reside in the subconscious, the most fundamental and impactful aspects of organizational culture are hidden from plain view: that is, they are not typically prominently displayed in inspirational posters on the walls.

There are many examples in the private sector where an organization's sense of purpose and its dominant culture have been the primary factor behind their long-term success or failure. A question of values and a sense of purpose that are critically important for the culture of human beings are equally important for the culture of an organization.

A definition of an organization's purpose creates fundamental assumptions that impact the metrics and the performance indicators that a company will use to make decisions. The story about Polaroid[48] may illustrate the idea:

> *As early as 1981, the company was making major strides in electronic imaging. By the end of the decade, Polaroid's digital sensors could capture quadruple the resolution of competitors' products. A high-quality prototype of a digital camera was ready in 1992, but the electronic-imaging team could not convince their colleagues to launch it until 1996. Despite earning awards for technical excellence, Polaroid's product floundered, as by then forty competitors had released their own digital cameras. Polaroid fell due to a faulty assumption. Within the company, there was widespread agreement that customers would always want hard copies of pictures, and key decision makers failed to question that assumption.*

Polaroid managers assumed that they were in the *photographs* business, when in fact the purpose of the company was to help customers preserve their *memories*. This fundamental understanding of purpose had a strong impact on the behavior of the organization. In 2001, Polaroid had its first bankruptcy.

A clear sense of purpose is connected to a measurable definition of value and is of greatest importance for an organization: a test of utility's "wisdom" should be its ability to manage its behaviors, actions, and decisions in ways that enhance their core values.

***A clear sense of purpose connects to a measurable definition of value and is of greatest importance for an organization.***

A key to success of human beings is their ability to create organizations that can accomplish far more than an individual could. Organizational models describe their structure, practices, processes, and cultures.

Throughout history, organizational models have evolved to adjust to the changing conditions and demands. An evolutionary path of organizational models has been proposed, as illustrated in Figure 5.3. Please note that the horizontal scale is logarithmic: the rate of change in organizational models has accelerated greatly in recent history. Every change brought about a new way for people to collaborate. It is important to note that some of the older models have not disappeared: several of these organizational models co-exist in today's complex world.

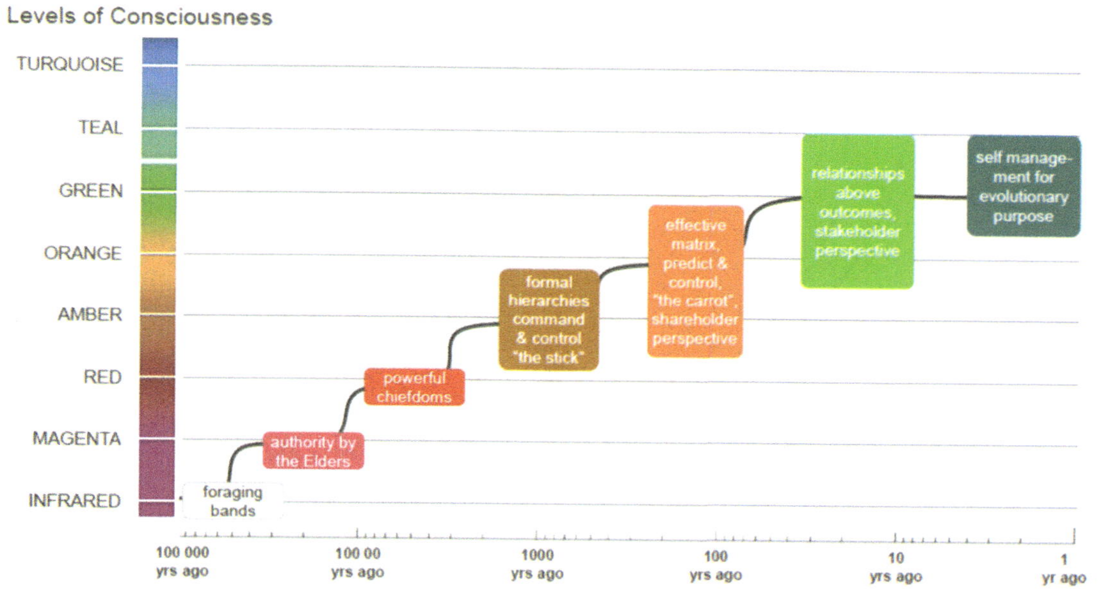


Figure 5.3. Evolution of Organizational Models: Change Factory (Laloux, 2014)

Figure 5.4 summarizes the key aspects of these organizational models. Note that each of the models/paradigms has been assigned a color: RED ("power, fear, chaos") is the color for organizations governed by chieftains, and the collaboration is based on fear: examples are mafia, street gangs, or tribal militias. For the purpose of this publication, we will focus on two most relevant organizational models in discussion about the water sector organizations: the AMBER ("conformist, hierarchy, stability") model and the ORANGE ("achiever, competition, profit") model. Table 5.2 shows a comparison between these two models.

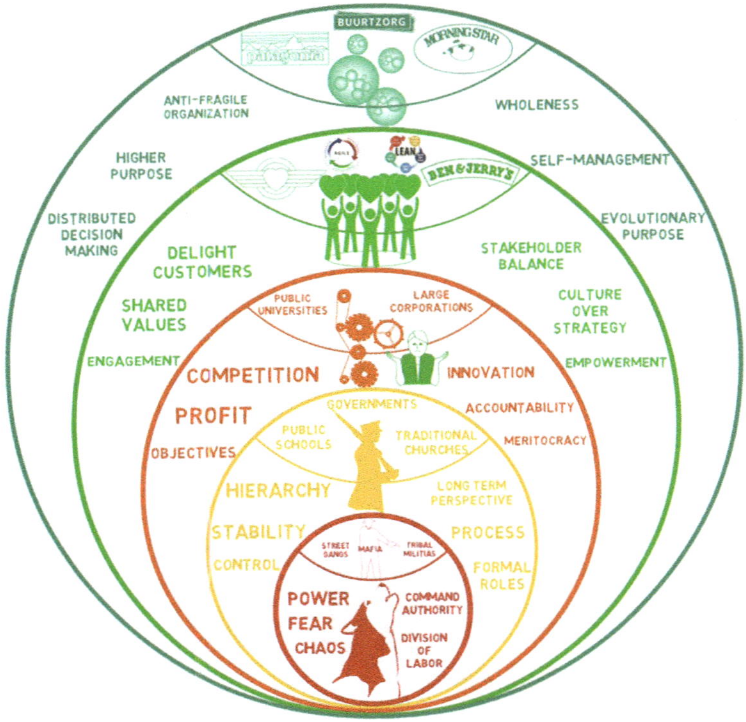


Figure 5.4. Dominant Models for Organizations (Laloux, 2014)

Table 5.2. Comparison Between Two Organizational Models (Laloux, 2014)

| **Conformist (Hierarchy—Stability) Model** | **Achiever (Competition—Profit) Model** |
|---|---|
| Metaphor | |
| Army | Machine |
| Characteristics | |
| Well-defined, formal roles | Competitiveness: profit and growth |
| Hierarchical pyramid | Innovation is key |
| Stability valued above all | Management by objectives |
| Rigorous | Top-down on WHAT; freedom on HOW |
| Key Breakthroughs | |
| Formal roles (stable hierarchy) | Accountability |
| Scalable | Meritocracy |
| Defined processes | Innovation (adaptability) |
| Examples | |
| Most government agencies | Large (e.g., multinational) companies |
| Military | Charter schools |

Early examples of the hierarchy—stability model include the Greek phalanx and the Roman legion; their strength came from clear roles, formal rules, defined processes and technologies, and a high degree of cohesiveness and structure in their organizational model.

The "competition—profit" model exhibits considerable variability and adaptive dynamics. Successful private companies typically go through several phases of evolution as they grow and adapt: they change to a different phase when their current organizational model no longer allows them to adequately address their business needs and challenges. Although they still operate within the same general organizational model paradigm (competition—profit), to grow and prosper they make adjustments in several aspects of governance and organizational structure and culture.

Each of the evolutionary phases of growth are described in Table 5.3 in terms of the management focus, organizational structure, top management style, control system, and management reward emphasis. The changes (from one phase to the next) are classified as "revolutions" that include substantial changes in each of these aspects. The changes are necessary for the company to grow and remain prosperous; they are driven by the need to remain competitive in terms of profitability and market share.

Table 5.3. Evolutionary Stages of Private Sector Organizations (Greiner, 1998[49])

| | Categories | | | | |
|---|---|---|---|---|---|
| | **Management Focus** | **Organizational Structure** | **Top-Management Style** | **Control System** | **Management Reward Emphasis** |
| **Phase 1** | Make and sell | Informal | Individualistic and entrepreneurial | Market results | Ownership |
| **Phase 2** | Efficiency of operations | Centralized and functional | Directive | Standards and cost centers | Salary and merit increases |
| **Phase 3** | Expansion of market | Decentralized and geographical | Delegative | Reports and profit centers | Individual bonus |
| **Phase 4** | Consolidation of organization | Line staff and product groups | Watchdog | Plans and investment centers | Profit sharing and stock options |
| **Phase 5** | Problem solving and innovation | Matrix of teams | Participative | Mutual goal setting | Team bonus |

The ability to change is a matter of survival: fewer than 12% of the companies that were on the Fortune 500 list in 1955 were still on the list in 2017 (American Enterprise Institute). The rate of change is increasing: a corporate longevity report from 2017[50] estimates that "at the current churn rate, about half of today's S&P (Standard & Poor) 500 firms will be replaced over the next 10 years."

In addition to evolutionary phases and the organizational dynamics within a single organizational model ("competition—profit") for private companies, there is also variability in organizational models themselves. Although this model is still dominant in large private companies, Figure 5.4 illustrates how additional organizational have recently emerged:

- The "shared value—pluralistic" where the applicable metaphor for this model is family. Leadership style is consensus-oriented and participative, and focus is on culture and empowerment to increase employee motivation. Examples are values-driven organizations like Ben & Jerry's (Ice cream, owned by Unilever, U.S.) and Southwest Airlines (U.S.).
- The "evolutionary, self-management" model has a living organism as its applicable metaphor. Leadership is distributed and driven by a strong sense of overall purpose. Examples include companies like Patagonia (Apparel, U.S.), FAVI (Metal manufacturing, France), and Buurtzorg (Health care, Netherlands).

Organizational culture includes many different aspects of interactions between individuals and the organization; aspects for individuals and for organizations are listed in Table 5.4.

Table 5.4. Aspects of Organizational Culture

| How Individuals: | How an Organization: |
|---|---|
| Make decisions | Defines/measures success |
| Take action | Defines rules/process for making decisions |
| Perceive and deal with conflict | Views conflict and disagreement |
| Reach agreement | Defines rules for power dynamics |
| Interact with other individuals | Defines the dominant mental model |
| Interact with superiors | Measures/rewards success of individuals |
| Interact with team members | Deals with failures |
| Interact with staff, direct reports | Perceives its own culture |
| Interact with clients, other business units | |
| Perceive organization's culture | |

Methods for assessing organizational culture of an organization generally fall into two categories: quantitative and qualitative. A quantitative method—the Competing Values Framework (Figure 5.5) has been used for thirty years and defines four types of organizational culture.

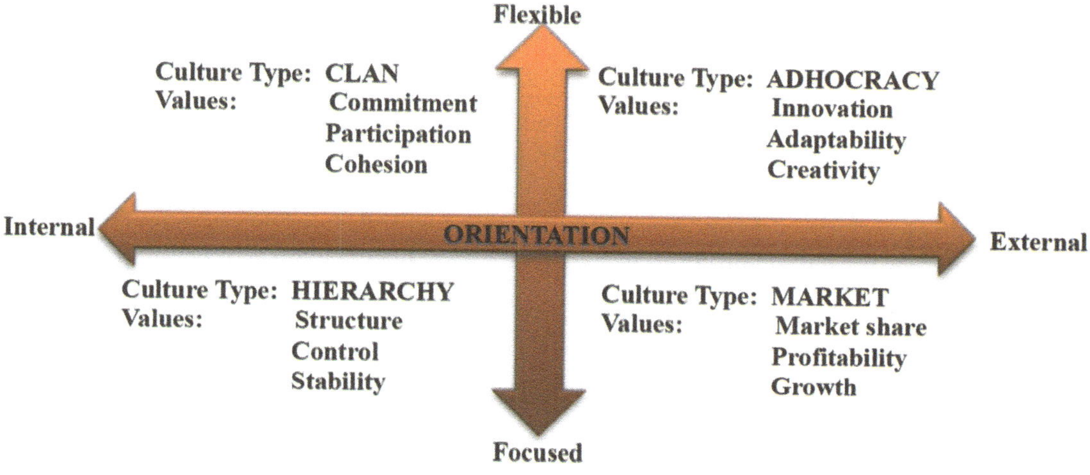


Figure 5.5. Competing Values Framework (adapted from Cameron & Quinn, 1999)

## 5.5 A Continuously Learning Organization—A Path to Wisdom

There are many internal challenges in the management of water sector utilities that should easily be solved using the rational thinking that is taught to students in engineering school. However, especially on planning, strategic, management, and policy levels, there are also a number of challenges that are more complex and fall into the category of "wicked problems."

A wicked problem[51] is a social, cultural, or organizational problem that is difficult or impossible to solve. The knowledge may be incomplete or contradictory. There are too many people and opinions involved, or the economy is a major obstacle. Often the problems are interconnected with other problems. Therefore, there is no apparent correct solution. Earlier we also suggested the term VUCA (volatile, uncertain, complex, and ambiguous) problems. Table 5.5 lists some characteristics of wicked problems.

Table 5.5. Characteristics of Conventional Versus Wicked Problems (see also Table 2.1) (McMillan & Overall, 2016)

| Characteristics | Conventional | Wicked (VUCA) |
|---|---|---|
| 1. Problems | Clear definition of problem, unknown solutions | No clear definition of problem—unknown and changing solutions |
| 2. Thought processes | Linear | Complex systems |
| 3. Time dimension | Task completed when problem solved | No time solution, politically determinate |
| 4. Nature of knowledge—expertise | Scientific solutions by experts | Problem definition is function of stakeholder views and perspectives |
| 5. Outcomes | Outcome is either true or false, successful or unsuccessful | Unknown outcome—may be better, worse, or acceptable. No "correct" solution. |
| 6. Problem approach | Scientific, knowledge protocols | Solutions are judgmental, depending on stakeholder views |
| 7. Problem characteristic | Loose coupling | Tight coupling |
| 8. Solutions characteristic | Cause and effect analysis | Multiple feedback analyses |
| 9. Value system | Shared values of outcomes | Values are in dispute, or in conflict |

***A wicked (VUCA) problem has no apparent correct solution.***

Dealing with complexity requires wisdom, and acquiring wisdom is an incremental process of integrating different types of knowledge, opinions, and interests into a holistic framework rather than mastering the details of a single aspect of knowledge. The four components of wisdom defined in Table 2.2 are equally applicable and important for an individual as they are for an organization. In both cases, achieving wisdom is a process of learning: a journey rather than the destination.

***Dealing with complexity requires wisdom.***
***Acquiring wisdom is an incremental process of integrating different types of knowledge, opinions, and interests into a holistic framework.***

Edgar Schein defined two categories of key challenges for an organization: internal integration and external adaptation.[52] Within each of these categories, there are many individual challenges of varying complexity ranging from relatively simple to very complex. The characteristics included in Table 5.5 and the results from the 2019 UAIM workshop (Figure 3.3) tell us that the most important challenges also tend to be more complex.

The outcome of the 2019 workshop (Figure 3.3) tells us that 89% of the challenges that utilities face when dealing with practical "bread and butter" work of a utility like asset management and/or delivery of projects have to do with the way the work is done (business processes) or the people who are included in getting the work done. For the around forty workshop attendees, it was apparent that this was a shared experience acknowledged by all as self-evident. There was also a general sense that "technology is the easier part" that can be fixed, but that the other issues require a different type of knowledge and capabilities.

## 5.6 Dealing with Wicked Problems

Most utilities have limited in-house resources and are typically constrained by the number and the type of staff that that they can hire and keep as employees. To get some of the work done, especially large capital projects, utilities typically need to turn to the consulting and vendor communities for help. External consultants can more easily provide highly specialized expertise in different subject matters because the cost of this (expensive) knowledge can be applied across a number of different clients. The experts that are provided by a consultant will have experience with doing the same type of work for other clients, while for internal staff this may be "their first rodeo." The method for engaging consultants is through contracts for professional services to execute a specific scope of work. As Table 5.5 indicates, such documents are more easily developed and executed to address the *conventional* category of problems.

Even for conventional problems, what has been discovered in the UAIM project, is that among teams of utility peers, there usually exists representative leading practices and processes. This expertise is shared for all to benefit; and challenges are also being addressed. This has minimized consultant involvement and has speeded enhancements for business processes.

When it comes to solving challenges that fall into the category of wicked problems, there is sometimes a temptation to follow the same method used for the conventional problems and engage an external entity. However, using the same approach to solve a significantly different problem may not yield optimal results. As Schein & Schein express in their classic text:

> *It is important to recognize that inside the organization there may be clear consensus on who has power, who has authority, and who has status, but this may be by far the most difficult element to decipher for someone who is not an insider.*

This does not mean that external consultants cannot play a meaningful and important role in addressing wicked problems that may require collaboration across the organization:

however, it does mean that the real "heavy lifting" to implement improvements will still most likely need to be done by management and staff within the organization itself.

To make this discussion less esoteric and more practical, let's look at a problem that at a first glance does not appear to be wicked at all: managing assets in a utility. A study by AWWA[53] included a survey of water sector utilities regarding their practices in asset management; their results are shown in Tables 5.6 and 5.7.

The results shown in Table 5.6 indicate that 90% of the utilities (in the United States) do not have the information about their assets organized in a well-defined asset hierarchy. Well organized, asset registry is the foundation for any analysis of assets: without a well-defined inventory/registry, we may not be able to answer even simple questions about the assets (e.g., how many emergency generators do we have?). In addition to not being well organized (i.e., not having a well-defined taxonomy), asset registry lacks important attributes and data.

Our decisions about managing assets will also require information about the condition of assets. Responses shown in Table 5.7 indicate that vast majority of water sector utilities in the United States do not use the information about the condition of their assets when they are making asset management decisions such as preventive maintenance, rehabilitation, or replacement of equipment.

Table 5.6. Survey Results: Asset Inventory/Register (AWWA, 2017)

| Which of the following describes your organization's asset inventory/asset register? | Yes Responses |
|---|---|
| Assets in the inventory/registry are organized as part of a well-defined asset hierarchy. | 10% |
| Advanced attribute data are largely populated for the assets in the inventory/register. | 12% |
| The inventory/registry contains more than 75% of assets. | 19% |
| There is a specific definition of assets versus nonassets for a majority of asset classes that governs the inventory/registry. | 14% |

Table 5.7. Survey Results: Condition Assessment (AWWA, 2017)

| Does the organization have a process in place to assess the condition of vertical assets (mechanical, electrical, HVAC*, and other asset types associated with facilities) and store the condition data in a spreadsheet or database? | Yes Responses |
|---|---|
| Condition assessment results are used to determine when long-term interventions should take place. Preventive maintenance is triggered based on condition rather than calendar intervals. | 9% |
| Condition assessment results are stored in a database for future analysis and trending. | 15% |
| Condition assessment is conducted on some noncritical assets in addition to critical assets. | 18% |
| Condition assessment is conducted on some critical assets to identify defects and trigger immediate intervention if necessary. | 21% |
| A formal process to assess the condition of vertical assets is developed. | 19% |

*Heating, ventilation, and air conditioning

According to a U.S. Congressional Budget Office report,[54] public spending on the water sector in U.S. utilities was $113 billion ($10^9$). Most of those funds are related to different aspects of asset management including development of asset management plans, designing new assets, operating and monitoring, and maintaining the assets. The fraction of costs for operation and maintenance (versus capital spending) has increased and now accounts for 60% of overall spending on U.S. public infrastructure.[55]

The technologies for asset management are readily available and include CMMS from several different technology providers. Such software systems provide the users with the capability to define the asset registry and to enter the attribute data. The underlying reasons for barriers to solving this challenge are not related to technology itself, but to people and many different business processes that are executed in an organization: the problem becomes wicked when people get involved.

Asset management is a type of challenge that water sector utilities typically struggle with:[56]

- It is defined in terms of processes that are executed by a number of different business units within a water sector utility.
- Different asset management business processes are supported by different IT systems that are purchased, owned, and operated by different business units.

Asset management is difficult not because technology is lacking, but primarily because it requires people from many different business units within the organization to work together: if the culture of the organization does not enable and reward collaboration, it is this culture that makes the barriers taller and transforms asset management into a wicked problem.

| |
|---|
| ***Asset management is difficult not because technology is lacking, but primarily because it requires people from many different business units within the organization to work together.*** |

# 6 UAIM Results—Models and Methods

This chapter includes some major results from the first three years of the UAIM effort.[57] Section 6.1 summarizes the first three years of the UAIM research toward WISE. The continued chapter describes different components of the WISE Framework. Business processes are key components in the value creation and have been the center of the UAIM results, as described in Section 6.2. In Section 6.3 an overview of the target of the WISE project and related research efforts are presented. The documentation of the results will be updated via a web portal on a continuous basis. In this way we aim to keep the document accessible, alive, and relevant.

## 6.1 The UAIM Effort Toward WISE

The UAIM effort started as a research project. However, from the very beginning, the intention has not been limited to producing a deliverable and then concluding the effort. Participants in UAIM have seen the project as a movement from the start: an effort that continues, grows, adapts, and improves. The intent of this document is to initiate a transition of the UAIM project into a movement with the expanded WISE Framework and the WISE project.

The mission of the WISE movement is to provide, sustain, and improve a platform that will enable the participants to collaborate on different aspects of utility management and improvement. In the immediate future (2020/2021) the participants will contribute to the WISE movement by:

- Developing business process models to rigorously document both their current ("as is") and future ("to be") processes and practices. Utilities will use common standard business process modeling notation to facilitate easier sharing of their models with other WISE participants.
- Conducting improvement initiatives within their utilities to implement improvements and share them with other participants through case studies published to the WISE platform.
- Researching issues related to the "people topics" including workforce, change management, and organizational culture, and creating guideline documents that provide assessment methods, maturity models, and management strategies for these topics/areas. The overall goal of these areas that include many "wicked problems" is to facilitate improvements and define practices that will help utilities achieve continuous learning.
- Provide continual learning opportunities, both from the platform, but also importantly, collaborating with peers.

Future collaborative efforts in the WISE movement may address other components of the WISE Framework such as technology and external impacts. This document includes placeholders for such future content that may be developed and included either by WISE or by other projects or initiatives that may decide to participate in WISE and contribute their content to have it integrated into the overall WISE platform.

At times, the journey to wisdom might require that we re-examine the "everybody knows" type of unspoken and entrenched rules that govern the behaviors in an organization, and step away from the well-beaten paths of the thinking patterns that have been created throughout an organization's history. The challenges of the future may differ from the challenges of the past. Organizations need to ensure that the learning process is embedded in their processes, technologies, people, and culture. Lack of organizational learning will limit our capabilities to deal with changes and reduce the possibilities for improvements in the future. Our journey will also be defined by the foundational qualities and values that will be our guiding lights: truth, trust, transparency, and respect for each other and nature.

***The journey to wisdom might require that we re-examine the "everybody knows" type of unspoken and entrenched rules that govern the behaviors in an organization.***

It is important to recognize that the participating utilities have been the principal creators of value during the UAIM project, and that this will remain true within WISE. The staff and management from the participating utilities have been developing and contributing the majority of the content, ideas, knowledge, and experience that has become part of UAIM deliverables and the artifacts that were shared and posted to the UAIM knowledge base.

Sometimes utilities may give more attention and credence to the voices of external parties, and the internal voices are not as readily heard. One of the major goals of the UAIM project has been to develop and strengthen a peer-to-peer network of staff and management and to provide them with an efficient mechanism to share their considerable knowledge and experience. This provides the management and staff from the participating utilities with contacts and shared artifacts that they can easily access when they are looking at problems and searching for solutions. Access to the collaborative efforts and shared knowledge may amplify the voice of each WISE participant within their own utility.

***We should develop and strengthen a peer-to-peer network of staff and management and to provide them with an efficient mechanism to share their considerable knowledge and experience.***

Plans are currently being developed by WEF to take an active role in the future of WISE and to incorporate it into its own programs. The transition from a "project" to an "initiative" will be more compatible with an idea of a movement: a dynamic effort without a predetermined end date that is continuous, growing, and adaptable.

## 6.2 UAIM Results—Business Processes

The business processes are the key components of value creation and have been frequently mentioned in the report. Consequently, this has been the focus of the UAIM research effort so far. Water sector utilities perform their functions by executing many business processes, including developing strategic plans and capital improvement programs; managing water quality laboratories, water, or wastewater treatment facilities; managing rates and decision processes, and so on.

### 6.2.1 Business Process Maturity Model

A most accurate and complete way to document process is by developing business process models. Business process modeling is the creation of graphical representations of business processes so that they can be analyzed and shared. A business process is a sequence of defined operations, tasks, or activities that can be executed by humans and/or machines. Business processes are put in place to help utilities reach multiple goals that generate value to rate payers, governments, communities, and the environment.

The capability maturity model includes five stages of business process capability/maturity, as illustrated in Figure 6.1: the key difference between different levels is in where the knowledge resides, and how it is applied:

- Initial: in individuals' heads. Training individuals is helpful.
- Managed: within teams. Team training and collaboration is helpful.
- Standardized: in accurate documentation about the processes, increasing the level of maturity from level 2. A system model is helpful.
- Predictable: the processes are managed quantitatively, based on some metrics. Monitoring and feedback will improve the processes.
- Innovating: the processes are managed through effective (and continuous) learning, leading to changes in management practices.

Business process modeling and management (BPM) is used to visualize and share knowledge of organizational processes, with the goal of improving alignment between people, process, and technology. BPM most notably can lead to benefits that help improve operational efficiency by increasing return on investment for projects as well as lowering total costs of ownership.

Once business processes are modeled, stakeholders gain insights into transformations that can optimize processes by reducing redundancy, lowering costs, improving practices, and adding appropriate technology to benefit the goals of a utility. Altogether, BPM is everything from analyzing to designing to simulating to executing business processes.

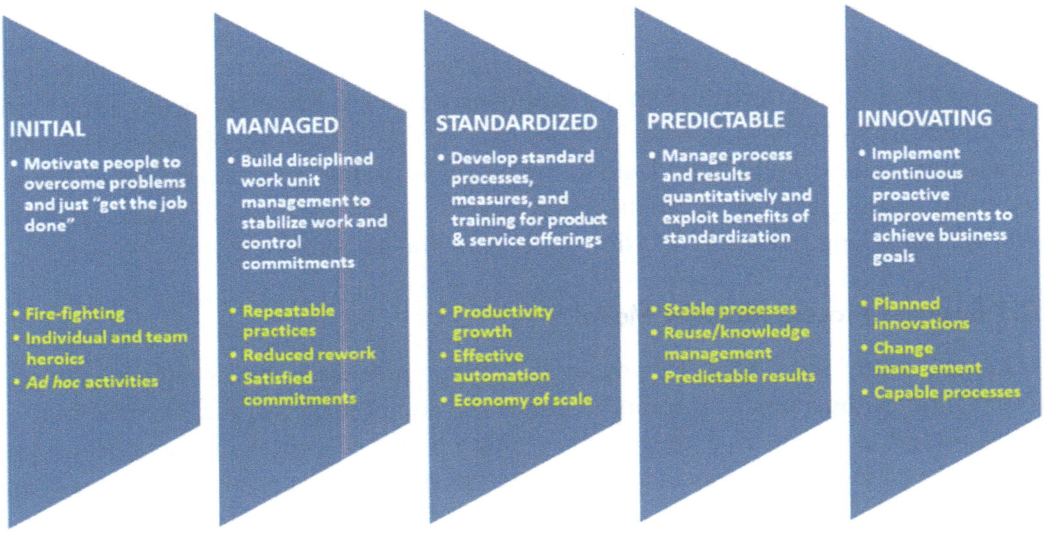


Figure 6.1. Process Capability Maturity Model (from Capability Maturity Model Integration).

A business process is always the responsibility of some participant; an organizational role, division, department, individual, or machine.

The activities, which are completed by process participants, are tasks that result in provision of a specific service or product for the organization.

The de facto standard for systematic modeling of business processes is the business process model and notation (BPMN). In the UAIM project, the BPMN is used to encourage completeness and consistency in business process transformation and improvement.

The information required in the BPMN enhances transparency and clarity in operational workflows, and answers questions like the following:

- What are the boundaries of the process, including all required elements and excluding others?
- Who is responsible for the process, and who participates in the process?
- What metrics are used to assess performance of the process?
- What triggers the start of a process, and when does it end?

- What artifacts, such as documents and data stored in a database, result from the process?
- Which activities must happen in sequence, and which can be performed in parallel?
- What decisions drive the process, and what are the business rules that must be applied?

### 6.2.2 Illustration of Continuous Improvement

The maturity model in Figure 6.1 defines a journey for the business process, starting at the "as is" assessment This is illustrated by the improvement loop in Figure 6.2.

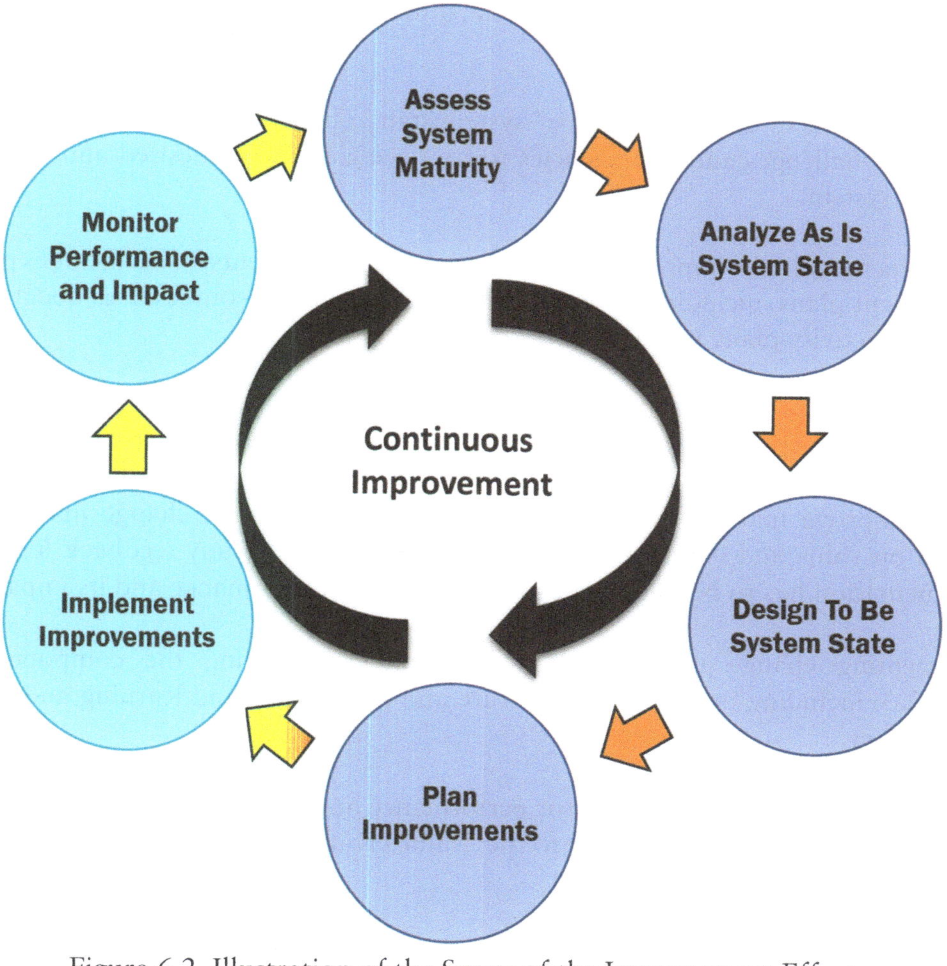


Figure 6.2. Illustration of the Scope of the Improvement Efforts
*Note*. The word "system" may refer to business processes, technology, or people. Here, we limit the discussion to business processes.

Any improvement has to start with assessing the maturity of the process. We need some model to explain how the system actually works. Such a system model is different for business processes, technology, and people. The model will tell:

- who is responsible for, and who participates in the process;
- what metrics are used to assess performance of the process;
- which activities must happen in sequence, and which can be done simultaneously;
- what technologies (including IT systems) are supporting the process;
- what decisions drive the process, and who makes them;
- what data are required to support decision making, and how that data can be accessed; and
- what artifacts (e.g., documents, data, information, drawings) are used within or generated by the process.

By now it is time to design the "to be" system state. The "as is" model can be adjusted to address challenges and inefficiencies to produce a model of desired and improved ("to be") system.

The next activity in the improvement loop is "plan improvements." This includes project management plans, including scope, schedule, budget, and resources. It is documented in various activity plans week by week for the near future.

Now it is time for "implement improvements." This is also demonstrated in the double-loop of Figure 2.3. This is a core mission of the WISE approach: to develop and apply a holistic analysis framework and a methodology based on systems modeling to help utilities improve maturity of their capabilities and implement change management focused on value and overall performance. Naturally—as in any feedback loop—the implementation has to be followed by monitoring the performance and its impact.

Implementing change management will require considering the components in Figure 2.3, including organizational culture and governance and learning and change management.

The last activity in the loop "monitor performance and impact" is the development of a "dashboard" (Figure 6.3), similar to the "value creation box" in Figure 2.3.

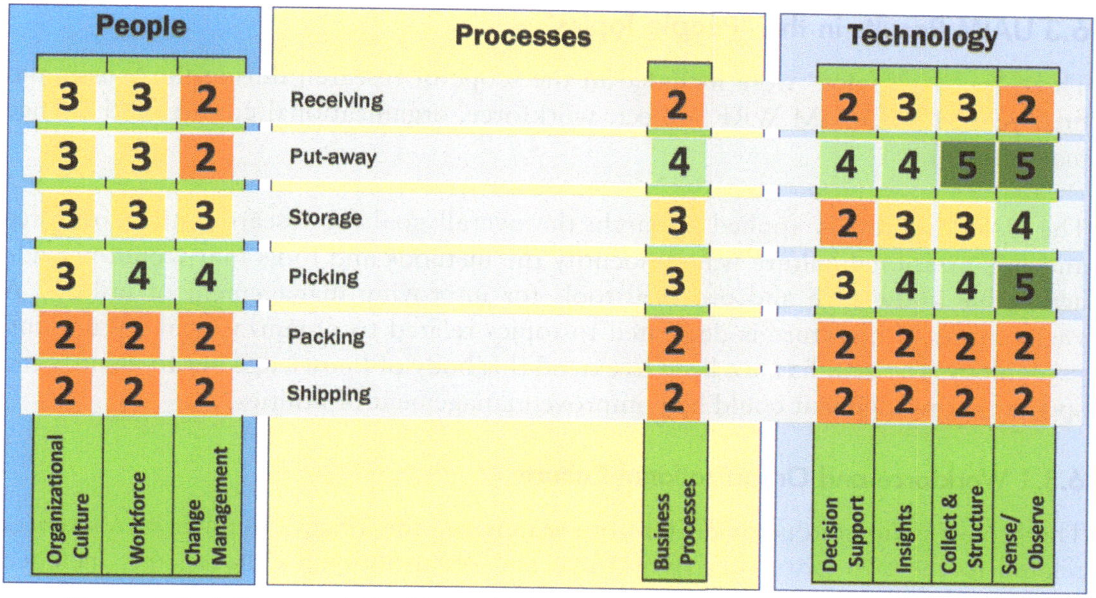


Figure 6.3. A Proposed Dashboard Monitoring the Performance of Different Categories of People, Technology, and Business Processes
*Note.* The numbers indicate the maturity level of the various aspects of value creation in the utility.

During the UAIM project, the maturity models and the assessment methods for workforce and organizational culture have gone through several cycles of proposals, reviews, and revisions. So far, the result is a Maturity Model v1.0, defining the aspirational goals and Assessment Method v1.0, defining the statements that would be used to determine the current position on the scale of capabilities for a specific case (organization). The details are documented in the UAIM Final report (2021).

The processes column includes several different business processes that are common in water sector utilities, and levels of maturity for all three components are indicated for each of the business processes. The significance of this is that (at least some of the) improvements may be achieved by applying the UAIM tools and methodology to specific processes. This step-by-step approach makes the methodology scalable and facilitates incremental learning.

The UAIM final report (2021) contains a number of case studies of BPM. Twelve utilities have submitted independent BPM, while four teams of utility partners have performed collaborative BPM and improvement results. The UAIM project team developed a business and decision modeling methods, tools, and guidelines document as a reference to help project participants understand better ways to apply modeling in business process improvement projects.

## 6.3 UAIM Results in the "People Topics"

Three "people topics" were included in the scope of research during the fourth and final year of the UAIM WRF project: workforce, organizational culture, and change management.

The UAIM project is applied research: the overall goal for research into workforce and organizational culture was to identify the methods and tools that would provide actionable knowledge and practical tools for improving management of utilities. A vast volume of literature is dedicated to topics related to organizational culture and psychology. The goal has not been to extend this body of knowledge but to identify the specific knowledge that could help improve management of utilities.

### 6.3.1 Workforce and Organizational Culture

The project team conducted a literature search and identified a number of references related to the two key issues (workforce and organizational culture); this included articles and books on different related topics including management, psychology, neuroscience, action science, and organizational culture. Several assessment methods were also examined—some of them proprietary and some of them in the public domain. Many of these sources had very valuable information; however, none of them provided a complete guide to solutions to the complex issues and challenges related to the people topics that would be readily applicable to the conditions that are prevalent in the public sector and in water sector utilities.

To ensure that the results of our research would be actionable and practical, the approach to analysis was linked directly to the methodology for making improvements. The effort started with the first step of the improvement methodology and the focus on maturity models and assessment methods for workforce and organizational culture.

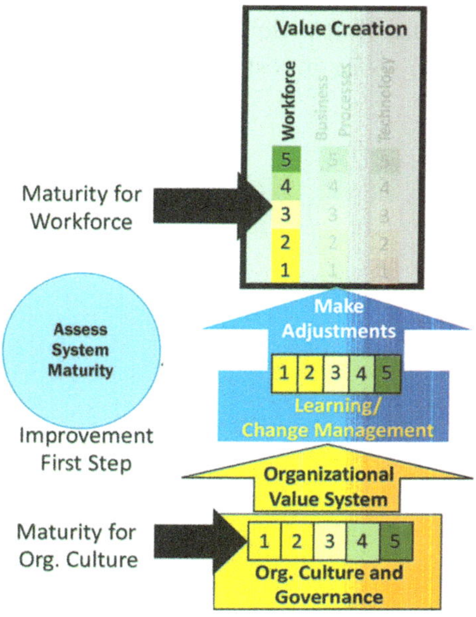


**From Figures 2.2 and 2.3**

UAIM participants recognized the complexities inherent in the people topics and agreed to the following:

a) To ensure that research stays grounded in the real needs and priorities of utilities, work would be led by utility partners and assisted by the project team.
b) Participants with specific expertise in these issues would be engaged.
c) The three topics are strongly connected through interactions and feedback loops. This required that research efforts also be connected.
d) Initial focus would be on developing maturity models and assessment methods.
e) Change management is defined as a process that integrates workforce and organizational culture into overall improvement efforts.

Through a number of meetings and workshops, utility partners collaboratively developed the draft (Version 1) of the assessment tool that corresponds to the maturity model categories. The key deliverable for the UAIM project was Version 1 of the assessment tool in the form of a tiered (nested) survey. As work will proceed through the WEF/WISE program, the plan is for utilities to conduct a pilot test of Version 1 in their organizations and improve the assessment method and tool based on the results from these pilot tests.

Assessments of workforce and organizational culture maturity will be important inputs to change management, since the "people issues" have a profound impact on an organization's ability to create value and/or implement change.

### 6.3.2 Change Management

UAIM project included the formation of a group made up from representatives of participating utilities with a shared interest in change management. This group held a number of working sessions, and utility partners shared the change management methods that are used by their organizations. Since change management is defined as a process, it was decided that the maturity models for business processes could be used for change management.

Since change management is an overarching process that "brings it all together" and may impact all of the components of value creation, a decision was made for UAIM to focus on workforce and organizational culture and that a detailed examination of change management would be moved to the future, to be investigated as part of the WISE program.

# 7 Transition from UAIM to WEF/WISE

This section will provide a high-level overview of the future work on the concepts and ideas that were started with the UAIM project and describe how this work will continue under the WEF/WISE program.

## 7.1 Technology in the Smart Utility Era

The challenges around technology are issues that have been around for a while but also new ones that have greater urgency, most specifically the issues around cybersecurity, data protection/privacy, and the need for a new breed of digital workers.

The ultimate value of third wave technology results will depend on overcoming these challenges and achieving success in several key areas.

| Successful Outcome Required | Challenges to Overcome |
|---|---|
| Providing robust cybersecurity while enhancing enterprise connectivity for increased enterprise resiliency to achieve heightened customer trust and satisfaction. | Utilities, like other organizations both in the private and public sector, are under the constant barrage of cybersecurity threats.<br><br>During the COVID crisis, we have seen the unprecedented and heightened need for access to information by a remote workforce to maintain utility operations and public trust. The COVID situation is unlikely to change anytime soon, so the challenges around data access, privacy, and security will be with us for years to come.<br><br>Addressing singular or multicloud security and standardization of cloud-based infrastructure will also be required. |
| Leveraging the integration of operational technology and IT platforms. | Over the past decade, we have seen the convergence of industrial process control technologies and IT systems. Traditionally, they have been separated. Without a new model for managing these systems, costs will become significantly higher, integration costlier, and innovation stymied.<br><br>A TOTEX (total expenditure) Model for technology will improve results and reduce costs. A key driver enforcing a TOTEX approach in water and wastewater utilities is the lack of complete and accurate engineering information captured during capital expenditure (CAPEX) that serves as feed for the operational expenditure (OPEX) cycle. Without such a model, utility will overpay for their technology investments, and the validity and accuracy of decision-making solutions will be brought into question. |

| | |
|---|---|
| Creating standardized and simplified technology solutions and platforms for standardized procedures that maximize knowledge retention and streamline processes to achieve operational efficiency. | Without a focus on data management, quality and changes in business processes, technology solutions will fail to achieve results and institutional knowledge will be lost.<br><br>Time to deliver new solutions must be focused on agile development so as results can be seen in months not years. Customers and employees want more immediate results. The days of delivering solutions in years rather than months is over. |
| Attracting and maintaining IT/ operational technology talent in the digital age could be the single biggest indicator of success. | Most utilities and government organizations still struggle to attract the talent required to deploy and maintain advanced technologies and systems. A skills gap exists, and it will become even more accentuated in the new era of digital workers.<br><br>Creative solutions will be required such as workforce training and retention, sharing a digital workforce among utilities (i.e., consortia) and outsourcing models. In addition, people coming out of universities and technical colleges will simply expect utilities to have the same tools they have at home at the workplace.<br><br>Without any changes in talent management, utilities and government will not be able to easily deploy and leverage technology. |
| Having a standard system and data architecture enables more effective and efficient system integration. | In many utilities, system and data architecture is seen as purely an IT issue versus a utility management business challenge. Although IT has a role in supporting these outcomes, solving these issues is essentially as important as financial rate planning and delivering capital improvement programs; hence, utility managers need to be thoroughly engaged and informed. Not doing so may very well result in under performance of daily business operations and costly failures. |

UAIM has not dealt systematically and thoroughly with the "technology" column. The role of technology has been documented within different business processes, which can define which systems are enabling these processes, what information is required for different activities, and so on. However, the technology column has not been examined in detail in the UAIM project, but the ongoing WRF project 5039 (Smart Utilities and Intelligent Water Systems) is further developing relevant results related to technology.

The scope of the WEF/WISE effort will expand to include the consideration of technology; the intention is to leverage the work that has been done as part of other initiatives conducted by WRF, WEF, and other organizations.

## 7.2 Integrating All Aspects of Value Creation and Change Management

The mission statement for the WEF/WISE program defines that the main goal of the program is

> *to develop and apply a holistic analysis framework and a methodology based on system modeling to help utilities improve maturity of their capabilities and implement change management focused on value and overall performance.*

The UAIM project provided some of the basic building blocks of this mission:

- The overall analytical framework based on creation of value,
- Improvement methodology,
- Maturity models, and
- Assessment methods.

UAIM has initiated a collaborative effort to examine change management as a key aspect of improvement.

Improvement should be driven by the organization's goals. Change management is defined as a process that is focused on coordinating different aspects of improvement activities including workforce, organizational culture, business processes, and technology.

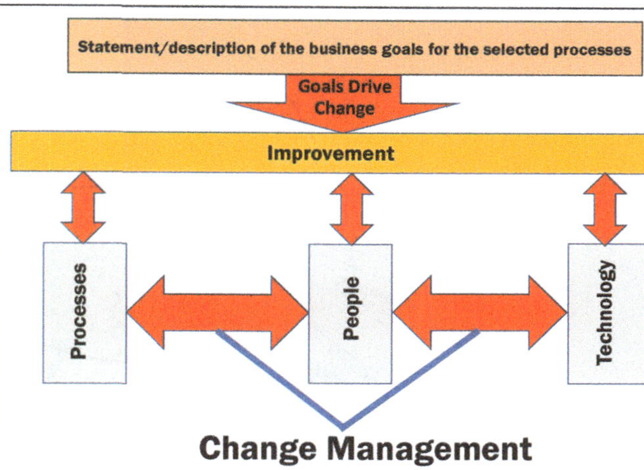


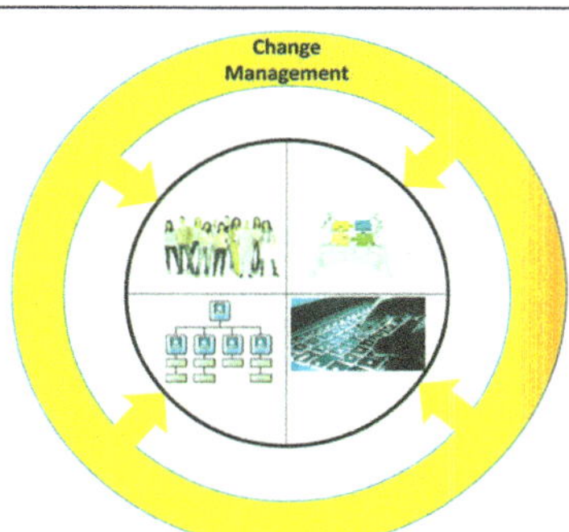


Change management therefore interacts with each component of value creation.

The first step of the improvement process is to assess the maturity level of the system. The scope of the UAIM project included the maturity models and assessment methods for workforce, organization, and processes, but not for technology. However, these items were included in the scope of the WRF 5039 (Smart Utilities and Intelligent Water Systems).

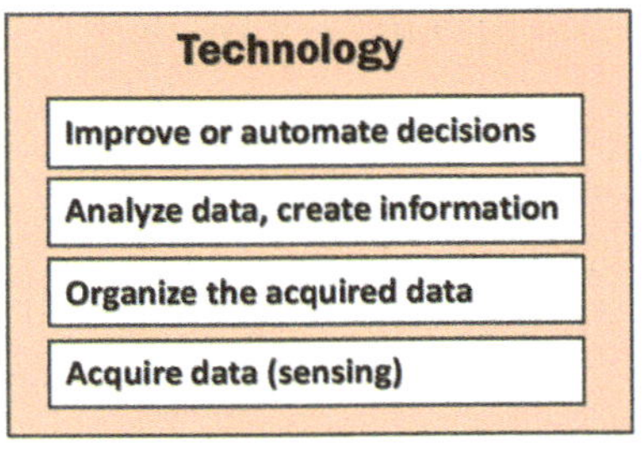


The WRF 5039 (Smart Utilities and Intelligent Water Systems) project proposed a maturity model that describes the capabilities of technology structured in four layers. The assessment method (to determine the capability/maturity levels) was also proposed and will be the starting point for considering technology within the WISE program.

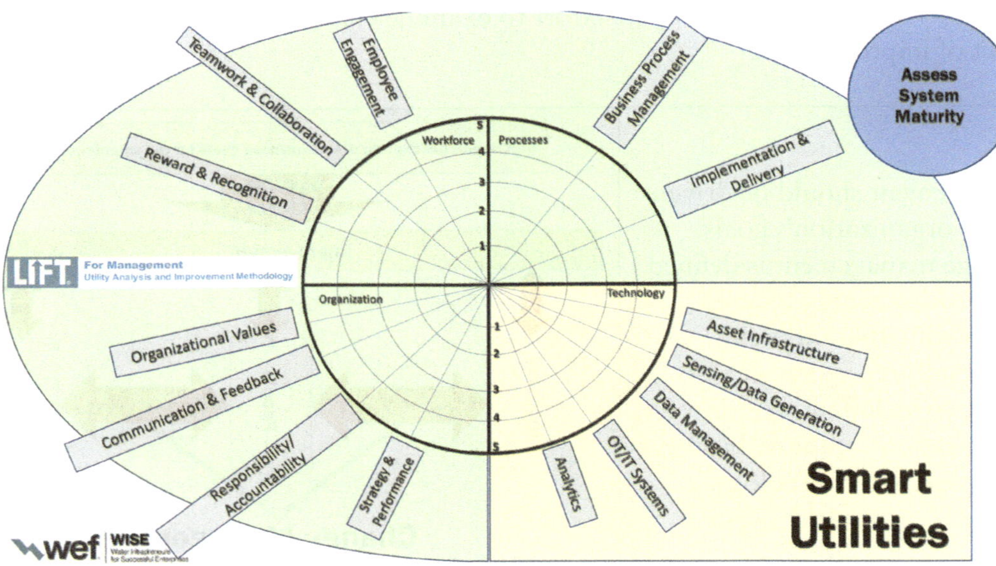


Figure 7.1. Assessment Methods for Workforce, Organization, Processes, and Technology from WRF Projects 4806 and 5039 to Be Leveraged by WISE

Figure 7.1 presents the research sources that will be used by the WISE program to construct an integrated and holistic view of capability/maturity levels across different aspects of value generation. The UAIM project produced the assessment methods for workforce, organization, and processes. The maturity models and assessment methods for technology are key deliverables for the WRF 5039 project smart utilities and intelligent water systems, and the results of this research will provide valuable information that will naturally fit into the "technology" column of the WISE Framework.

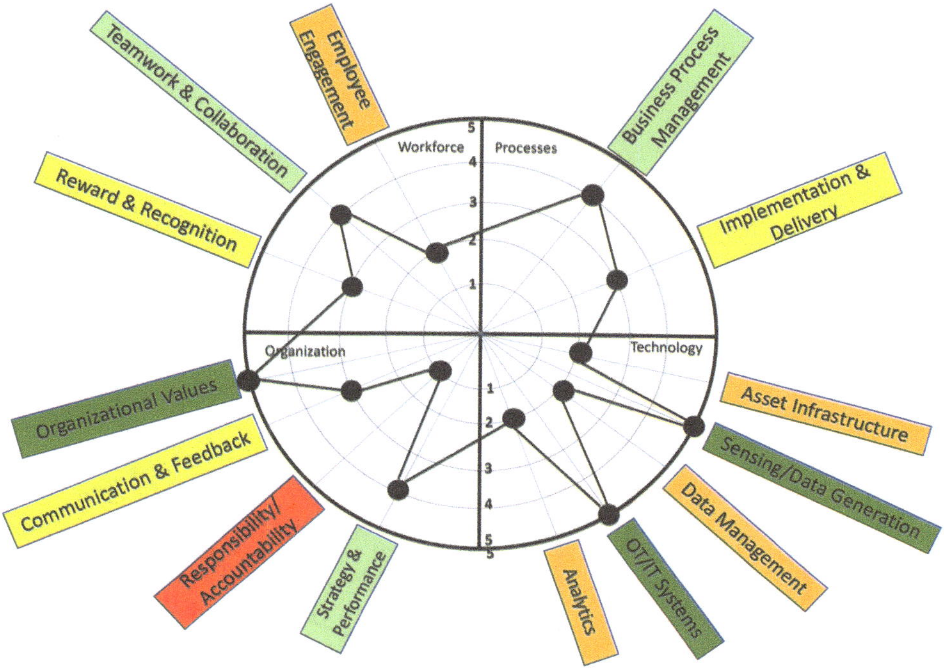


Figure 7.2a. Results of Assessment

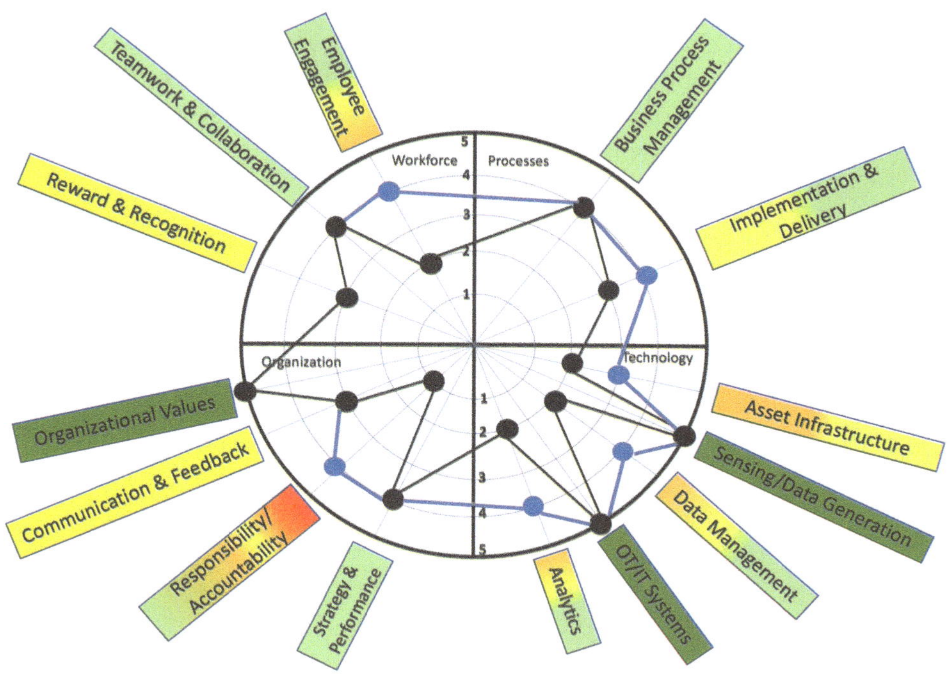


Figure 7.2b. Goals for Improvement (Blue Dots)

With different components, different maturity models with each component, and different assessment methods that structure different capabilities into categories, things can get quite complicated. Figure 7.2 provides an example view (potentially a dashboard) of the "as is" state of the system and the desired "to be" state of the system that includes all four components (workforce, organizational culture, process, and technology) and the capabilities associated with each component. Such a view summarizes the results of assessments and the scope of the planned improvements.

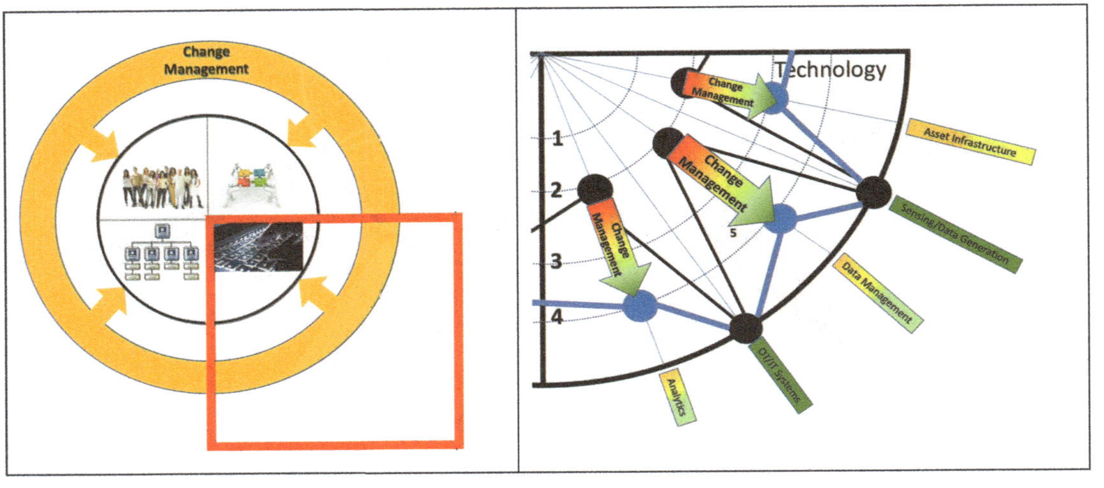


Using technology as an example, the figure on the right shows how the results from maturity assessments can be used to define the scope of improvement and change management—in this case for technology but the same principles apply to the other components—organization, workforce, and processes.

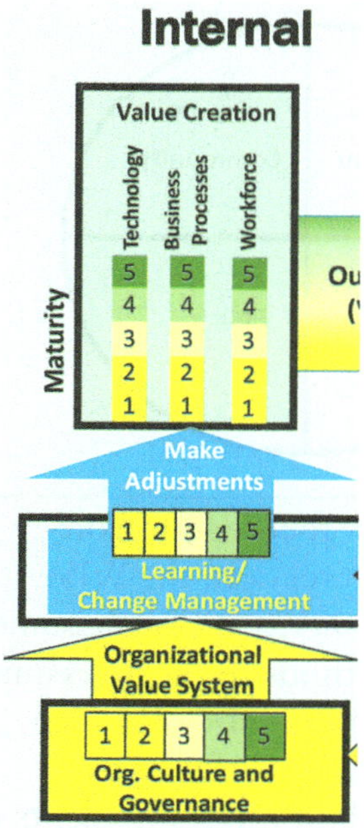


From Figures 2.2 and 2.3

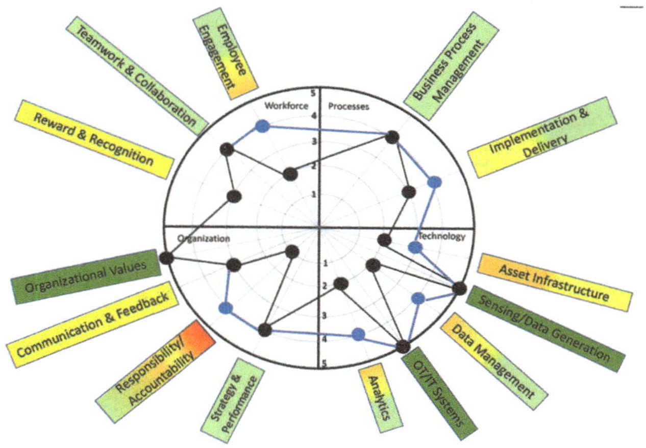


It is important to note that the models and methods discussed so far in this chapter focus on the creation of value internally within the organization. Therefore, this addresses only one part of the overall WISE Framework, and it does not take into consideration the impact that the organization has on the external world.

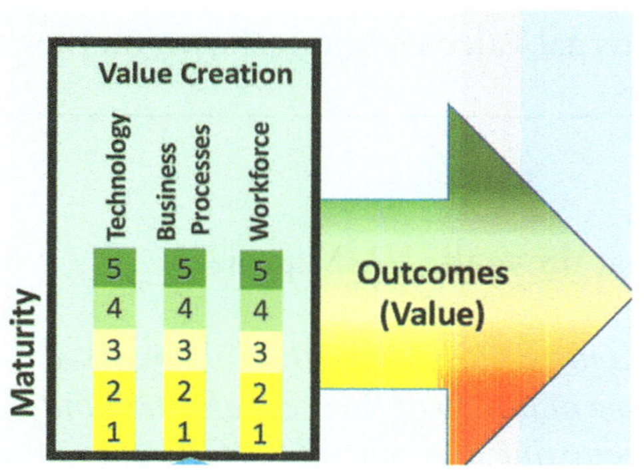


From Figure 2.2

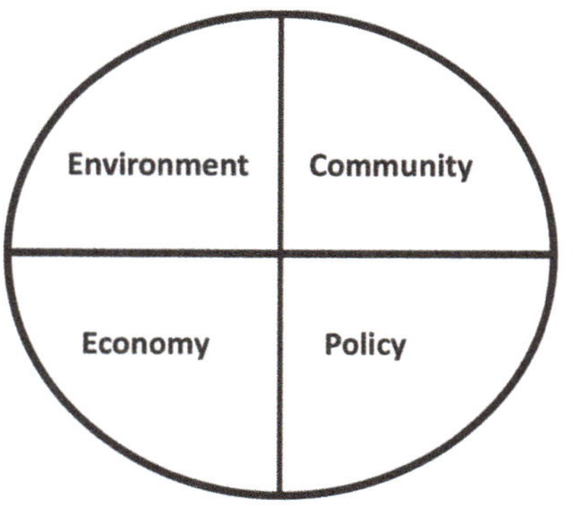


Each organization produces outcomes that deliver different components of value to the outside world: creation of value ***within the organization*** can be viewed in terms capabilities, expressed in terms of levels of maturity in the management of workforce, business processes, and technology. The ***quality of impact on the external world*** can also be described in terms of different levels of maturity related to different components of impact.

In the current/initial WISE model of the external world, it is proposed that the impact includes several components including the local economy, community, compliance with policies and regulations, and the environment. Each of these components introduces different, and sometimes competing, demands on the organization.

<table>
<tr>
<td>
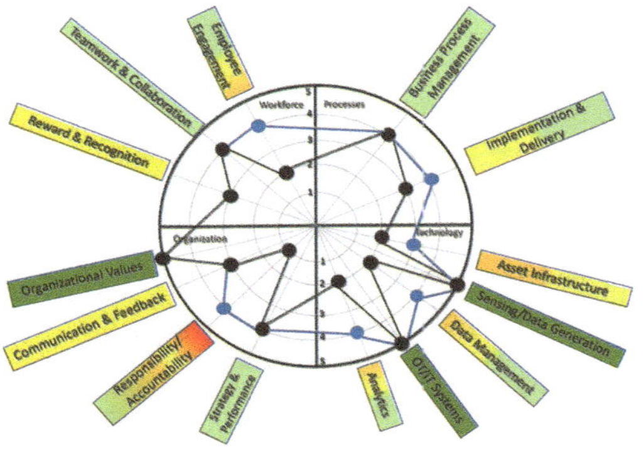

</td>
<td>
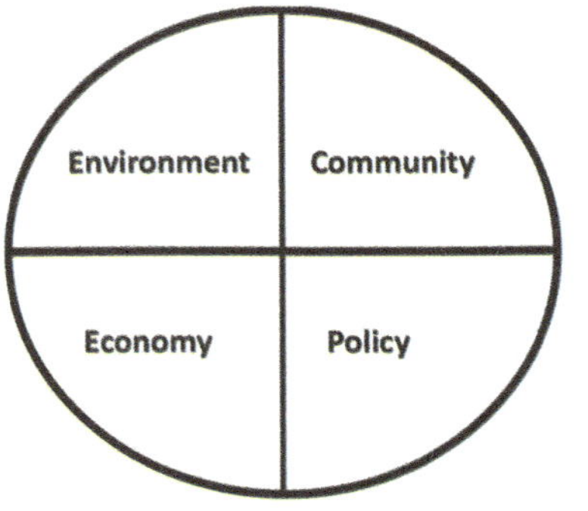

</td>
</tr>
<tr>
<td>We used the “spider diagram” to assess the current (“as is”) maturity of value creation within the organization (black dots) and to define the desired (“to be”) state of each of the four major components (blue dots). Each of the components was further subdivided into categories of capabilities, so the diagram illustrates the specific capabilities that need improvement.</td>
<td>The UAIM project focused only on the internal value creation; the WISE program will expand to apply the same principles and methods to the assessment of impact on the external world. The goal for WISE will be to define the categories for each of the components of impact and develop models and methods to produce a dashboard for impact that is equivalent to the dashboard for internal value creation (shown on the left).</td>
</tr>
</table>

## 7.3 A Project Becomes a Movement

The mission of WISE retains the basic ideas that drove the UAIM project:

> *To develop and apply a holistic analysis framework and a methodology based on system modeling to help utilities improve maturity of their capabilities and implement change management focused on value and overall performance.*

In 2017, when the work to accomplish this mission was just starting, the effort was defined and structured as a research project (initially sponsored by WE&RF, later continued by WRF). The research project framework provided some advantages in the early stages:

1. Most utilities are comfortable with the idea/framework of a project because it provides a well-defined scope, schedule, and budget. The framework for executing projects is very familiar to them.
2. Sponsorship by an established national research organization includes transparency and oversight by the Project Advisory Committee.

3. Although the mission included the mentions of "holistic analysis" and "system modeling" from the very start, in the very beginning it was necessary to begin by identifying individual components of the "system" and to start addressing them individually first. Our choice was to start with business processes (rather than people or technology).

As we are now transitioning from UAIM to WISE, some of the initial constraints have gradually dissolved: everybody engaged in UAIM was learning by doing, and gradually participants naturally adopted a somewhat different perspective. An awareness emerged that the "project" framework is not optimal for WISE:

1. This effort is quite ambitious: managing a water sector utility is very complex. As one of the UAIM utility partners Jackie Jarrell said, "this is really innovation for management." Management is a wicked problem that does not have a single best solution that can be documented in a report at the end of a project.
2. Many challenges in managing utilities are the result of fragmentation: breaking down the problem into different pieces and solving each piece separately. A holistic approach requires a paradigm that considers the interactions between the pieces: a methodology based on system thinking that includes iterative improvement. WISE is about continuous learning and evolving improvement, not about a delivery of a discrete product that will be finalized.

It is better to think of WISE as a movement that will grow organically and adjust as our knowledge matures. Future efforts of WISE will require engagement of people with different skills, talents, backgrounds, and knowledge. We have already seen this in the last year of UAIM when our scope was broadened to include workforce and organizational culture. While the first three years of UAIM mostly engaged people with engineering background, in the last year of UAIM the team expanded to include professionals from human resource departments and senior management.

The term "movement" describes a dynamic process that engages people to collaborate on a common goal, guided by a common vision, and this is exactly what we want WISE to be.

## 7.4 Next Steps

As shown in Figure 7.1, the WISE program will continue the research that has been conducted on two WRF projects: UAIM (WRF 4806) and Smart Utilities and Intelligent Water Systems (WRF 5039). WISE will expand the use of the improvement methodology that was introduced by the UAIM project and continue to apply the methodology steps to different areas. Table 7.1 shows efforts initiated and conducted by the two WRF projects and the areas that will be addressed by the WISE program.

Table 7.1 Scope for WISE

| | Value Creation Within the Organization | | | External Impact |
|---|---|---|---|---|
| UAIM Steps of Improvement | Business Processes | Organization and Workforce | Technology | |
| Assess System Maturity | UAIM | UAIM | WRF 5039 | WISE |
| Analyze System State (maturity) | UAIM | UAIM/WISE | WRF 5039 | WISE |
| Design to Be Improved State | UAIM | WISE | WISE | WISE |
| Plan Improvements | UAIM | WISE | WISE | WISE |
| Implement Improvements | WISE | WISE | WISE | WISE |
| Monitor Performance | WISE | WISE | WISE | WISE |

The WISE program will also continue the work on change management that was initiated by the UAIM project. WISE considers change management to be a process that integrates the components of value creation within an organization.

The WISE program will officially start on September 1, 2021. At the time of this writing, WEF, the utilities participating in WISE, and the WISE project team are preparing the work plan for the first (2021–2022) year of the WISE program. The activities will be phased to address the strategic and tactical priorities of the participating utilities.

# 8 LEAP—Leaders for Emerging Applied Practices

Both UAIM and WISE are applied research efforts focused on a structured methodology for improvement of water sector utilities. The approach is tailored to water sector utilities as they are today and considers traditions and constraints common to such organizations; the emphasis is on improvements that could be achieved in the relatively short term and have an incremental impact. As authors were finalizing the manuscript, concerns were raised that key global sustainability challenges may require a more transformative approach that would reshape the dominant paradigms in water sector utilities. This new effort, Leaders for Emerging Applied Practices (LEAP), would be conducted independently and concurrently with WISE.

Communication, collaboration, and exchange of ideas and artifacts between LEAP and WISE will be encouraged; however, the two initiatives will be administrated separately so that one does not impede or constrain the other.

The focus of WISE is primarily on how to perform better and achieve the mission of water sector utilities as it is commonly defined today. LEAP takes an explorative and piloting approach. We ask what is the highest future potential we see, and how can we move in that direction.

Referring to the figure resulting from the UAIM research, we see that while WISE has its focal point on maturity and value creation in regards to technology, business processes, and workforce, the LEAP value creation box puts emphasis more on the external world, that is, community, economy, policy, and environment, and its impact on individual development within the utility (Figure 8.1).

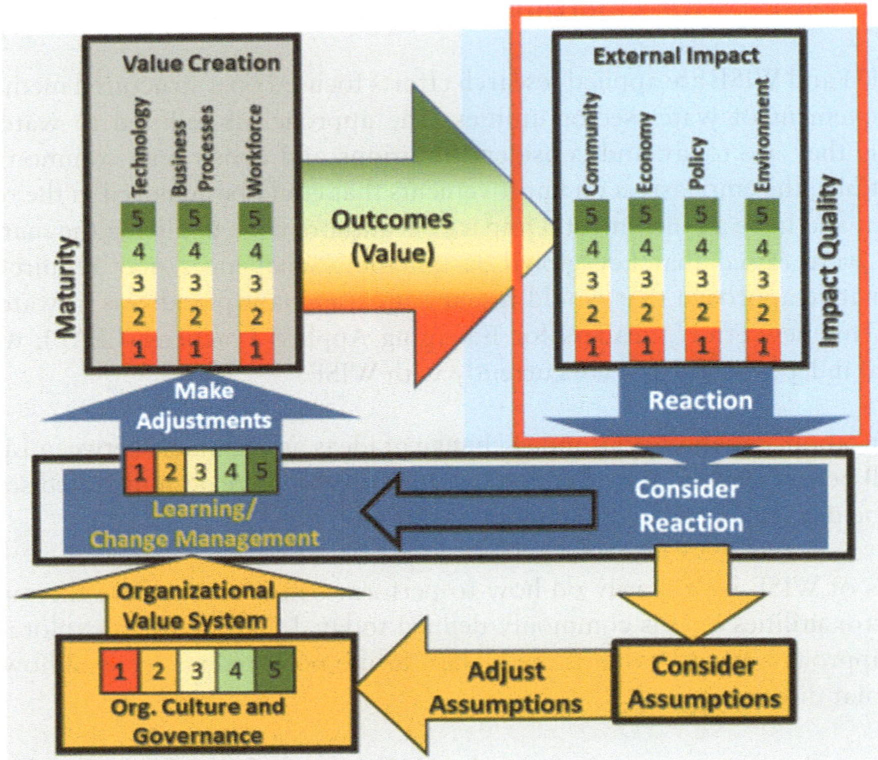


Figure 8.1. External Impact Component Within the WISE Framework.

In the UAIM framework, LEAP is focused on addressing the "external impact" box as a focal point, whereas WISE focuses more on the internal "value creation." Another difference is that, in the feedback loop, WISE focuses on organizational culture and governance, whereas LEAP has a special focus on individual's developmental processes and their ability to effect change.

This section presents initial thoughts about the LEAP initiative; however, this new program is just beginning, and the specifics will be refined collaboratively by the participating utilities after the LEAP group has been formed. By working with groups of innovative professionals from utilities from around the world, LEAP will explore and experiment with emerging practices for co-creating a new future of sustainable and regenerative stewardship within water.

It ought to be emphasized that the LEAP effort should be financed completely separately from the WISE project. The aim is that any results of the LEAP project should be freely and openly available to all utilities.

## 8.1 The Purpose of LEAP

In the prologue, we outlined some of today's major water challenge: to meet human water needs in an equitable way without surpassing the planetary boundaries such as climate change, depletion of fresh water resources, and disturbing the nitrogen and phosphorous cycles while reestablishing healthy conditions for a thriving biosphere. This is what is required to have both human systems and ecosystems flourish in the future. Many of the systems that we stress and disturb now may not be possible to keep for the future but can be eternally lost.

The world has yet to establish its first truly sustainable water system operating in unison with its people, the watershed, the atmosphere, and its local ecology. To succeed with that requires a radical change in our utilities as well as in our communities. This reshaping of the water system will include new innovative technology as well as organization and must happen by seamlessly transforming the current system and keeping the systems unstopped in operation—radical change through many small steps (Figure 8.2).

Figure 8.2. There Is a Major Transformational Challenge to Reshape Our Current Water Systems to the Future of Fully Sustainable Water Systems

Water utilities can play a central role in this work of vision and transformation if we increasingly think in terms of water stewardship. This enables professionals in the water sector to take a role as facilitators for this transformation.

To think like a water steward entails:

- Working from an inner sense of vision and responsibility,
- Daring to work with complex issues,
- Experimenting and learning by doing,
- Moving ahead consistently in many small steps,
- Taking in feedback continuously from the work—at all organizational levels,
- Being firmly grounded in science,
- Being soundly grounded in emotional intelligence,
- Developing skills to work co-creatively,

- Developing skills to facilitate collaboration,
- Being able to communicate vision, and
- Thinking seven generations ahead.

All leadership is founded in oneself. We need to learn to lead ourselves before we can lead others. LEAP will provide a platform to experiment with this challenge.

We envisage the platform will consist of a number of LEAP groups (5–20) from different utilities around the world. The groups will represent different types of organizations and cultures. Each LEAP group is challenged to start a movement toward sustainable water practices and stewardship within their organization. Meeting such a challenge requires a certain measure of inner motivation, which can only be nurtured where there is room for reflection, experimentation, and planning relevant actions. LEAP provides such a process that supports the development of each individual in the group inspired by new emerging leadership practices. The process begins by asking three key questions:

- How are you feeling?
- What is keeping you busy?
- What are you going to do about it?

LEAP will work as a "laboratory" or an "incubator" for new thinking, new solutions, and new ways of collaboration—a new kind of leadership.

The three key activities in the LEAP initiative are:

1. *Testing and experimenting with emerging advanced practices:*
   A number of advanced practices to support and facilitate the sustainable transformation are currently being developed at various institutes and groups around the world. A key activity in the LEAP initiative is to look for these new methods for inspiration and find ways to transform these emerging practices into something functional in the life of water utilities.
2. *Making new methods and principles freely and openly available to all utilities:*
   Through the work with the new practices, they are transformed to fit into the domain of water. The LEAP initiative will document and share the results in the form of practices, processes, and principles for other utilities to utilize or continue building on top of them.
3. *Work with own intrinsic motivation for sustainability:*
   A key shift is to work from intrinsic motivation of what we feel/are called to do rather than struggling in the existing control/and/command mindset. This does not mean that the group will work on random issues depending on personal interests. Taking on a public service role means understanding the specific needs of the watershed, including both the people and nature, and

engineering good solutions that serve both. Understanding the necessary transition to a sustainable utility future in ourselves and shifting from extrinsic to intrinsic motivation is a key factor to succeed in the journey. To grow and find balance in oneself while being at service for the greater good is a great challenge, and it requires practice as well as new habits of thinking and acting.

Outside of mainstream management culture, new ideas and experiments are emerging. These methods focus in new ideas for human development working from a concept of "living systems"—as a contrast to the now classical industrial frame of mind. These represent a body of knowledge, insights, and practices about a new way of leading and transforming. On the LEAP platform, we will work with adapting these methods to the context of the water sector. We believe that these new ideas and methods can reshape the future of sustainable water systems.

Examples of new principles are:

*SDGs*

The SDGs[58] were presented by the United Nations as a first ever method to think holistically for the whole world about the challenges ahead. The goals cover 17 areas including social, environmental, and economic goals: no poverty, zero hunger, good health and well-being, quality education, gender equality, clean water and sanitation, affordable and clean energy, decent work and economic growth, industry, innovation and infrastructure, reduced inequalities, sustainable cities and communities, responsible production and consumption, climate action, life below water, life on land, peace, justice, and strong institutions and partnerships for the goals.

*Alliance for Water Stewardship*

The Alliance for Water Stewardship[59] provides a framework for certification of water stewardship mapping, measurements, and practices. Stewardship means taking care of something that we do not own, in this case water, which is a critical commons, that is common resource. The certification ensures a transparent system for ensuring a socially and culturally equitable, environmentally sustainable, and economically beneficial use of water by including all relevant stakeholders.

*Doughnut Economy*

Our economic thinking seems to repeatedly come in the way of doing what we believe to be the "right action." The origin of the word "economic" is the Greek word for household management. Doughnut economy[60] suggests a new principle that takes the idea of "household management" on a global level serious. In the picture of the "doughnut," the outer rim of the doughnut signifies the planetary boundaries, which provide the ceiling for our actions. The inner rim of the doughnut is defined by the social boundaries, for example, food security, health, and education, and provides the social foundation. If our planetary household management overshoots on the environmental boundaries or falls short on the social foundation, we depart from the area of sustainable operation—either ecologically or socially.

*Biomimicry*

Biomimicry is the idea of mimicking the solutions found in nature. The idea is that nature has had an innovation process spanning billions of years of evolution. Succeeding in learning from nature in design challenges means standing on the shoulders of this very robust system that nature has "developed." Aquaporin, biomimetic membranes, and seaweed based flocculants are examples of biomimicry methods applied in water treatment.

*The Presencing Institute*

The Presencing Institute[61] has provided a systematic way of working and thinking of leadership in the transition to this new era.[62] The work is centered around a U-shaped process by which the participants "learn from the emerging future" as opposed to keep copying or downloading the past. The practices facilitates a method of "open mind," "open heart," and "open will" to reach a point at the bottom of the U, where new ideas for solutions emerge. These ideas are then crystallized, and prototypes are tested before making the solutions widely available for easy implementation.

Additionally, a number of thinkers have made elaborate contributions on the insights into this transformation, including people such as

- James Lovelock (*The Gaia Theory*[63]),
- E. F. Schumacher (*Small Is Beautiful*[64]),
- Arne Naess (*Deep Ecology*[65]),
- Clarissa Pinkola Estes (*A New Vision for the Feminine*[66]),
- Charles Eisenstein (*The More Beautiful World Our Heart Knows Is Possible*[67]),
- Michael Stubberup and Steen Hildebrandt (*Leadership from the Heart*[68]),
- Carol Sanford (*The Regenerative Business*[69]),
- Thomas Hübl (*Healing Trauma*[70]),
- R. Buckminster Fuller (*Operating Manual for Spaceship Earth*[71]),
- Vladimir Vernadsky (*The Biosphere*[72]),
- Stephen W. Porges (*Polyvagal Theory*[73]),
- Jacob Stam (*Constellation*[74]),
- David Sedlak (*Water* 4.0[75]), and
- many more.

The LEAP platform apply the ideas of this domain to work with what might be, creating clear vision and experimentation.

## 8.2 Why Join LEAP

The LEAP initiative invites utilities ready to develop a new sense of sustainable leadership. Reasons to join the platform include:

- Start a movement in the utility toward sustainability and water stewardship;
- Learn new sustainability practices and practice to apply them;
- Personal growth and development of leadership capacity;
- Work with existing projects—and raise the bar;
- Get a global perspective and be inspired to think globally;
- Have a local "think tank" to inspire future projects in the utility; and
- Have a good collaborative experience for all involved.

So many of us are yearning for a change for the better in our practice of work. We yearn for the opportunity to change our world into a sustainable one through a wise collaborating community. We want to be those who rose to the challenge and who trusted each other enough to change our culture, economy, and technology. We yearn in sum to be good ancestors.

## Abbreviations

| | |
|---|---|
| AMI | Automatic Meter Information System |
| AMR | Automatic Meter Reading System |
| AWWA | American Water Works Association, https://www.awwa.org/ |
| BPM | Business Process Modeling & Management |
| BPMN | Business Process Model and Notation |
| CADD | Computer Aided Design and Drafting System |
| CAPEX | Capital Expenditure |
| CIP | Capital Improvement Program |
| CMMS | Computerized Maintenance Management Systems |
| DCS | Distributed Control Systems |
| GIS | Geographic Information Systems |
| GWRS | Groundwater Replenishment System |
| IWA | International Water Association, https://iwa-network.org/ |
| IWRM | Integrated Water Resources Management |
| LEAP | Leaders for Emerging Applied Practices |
| LIFT | Leaders Innovation Forum for Technology, https://www.waterrf.org/utility-management-lift |
| LIMS | Laboratory Information Management Systems |
| OCSD | Orange County Sanitation District, https://www.ocsd.com/ |
| OCWD | Orange County Water District, https://www.ocwd.com/ |
| OPEX | Operational Expenditure |
| SCADA | Supervisory Control and Data Acquisition |
| SDG | Sustainable Development Goals |
| TOTEX | Total Expenditure |
| UAIM | Utility Analysis and Improvement Methodology |
| UMC | Utility Management Conference |
| VUCA | Volatile, Uncertain, Complex, Ambiguous |
| WE&RF | Water Environment and Reuse Foundation, https://www.waterrf.org/ |
| WEF | Water Environment Federation, https://www.wef.org/ |
| WEFTEC | Water Environment Federation's Technical Exhibition and Conference |
| WRF | Water Research Foundation, https://www.waterrf.org/ |
| WSVM | Water Sector Value Model |

# Acknowledgments

## Dedication—John F. Andrews (1930–2011)

Gustaf and Cello would fondly remember the man who had introduced us some forty years ago at the University of Houston, Texas: Cello's Ph.D. advisor and Gustaf's friend and collaborator Professor John F. Andrews. We would remember his generosity, unwillingness to "play politics," and his sense of humor. Already in the 1970s he was one of the true driving forces to increase the collaboration between WEF and IAWPR (the predecessor of IWA).

*Gustaf's story*: Having a control background, I wanted to apply control methods for wastewater treatment systems in the early 1970s. "Traditional" civil engineers told me that "control can solve nothing." It was obvious that nobody is a prophet at home. I spent a sabbatical semester in 1975 with John Andrews in Houston, learning more about the treatment processes from John and sharing control knowledge with him. A Swedish group of wastewater researchers came to visit John and wanted to interview him about control and operation. Being such a generous person, he immediately told them: "I know nothing about control, please ask Gustaf. He is here." That moment became a turning point for my recognition at home. The competence seems to increase with the distance from home. John became my great mentor for many years, and I enjoyed his friendship until his death in 2011.

*Cello's story*: I arrived in Houston in 1980 to start graduate school under Professor Andrews. His secretary told me to go right in—I would not need an appointment. John Andrews was sitting with his feet on his desk that had several high stacks of papers, with reading glasses half down his nose. I timidly introduced myself, and John jumped up and greeted enthusiastically. He said, "I want to show you around and introduce you to some people," and then he looked down at his stockinged feet. He reached into a drawer, pulled out a pair of stockings that had shoes painted on them, and put them over his regular stockings. Then he looked happily into my stunned expression and said, "Now we are ready to go!"

I would remember John wagging his finger and saying "Interactions, Cello, remember that it is all about interactions." One never stopped being an Andrews student: he kept sending us "must read" papers years after we had graduated and after he had retired to a "mountain top in Arkansas." Yes, John, I finally get what you meant: it indeed is all about the interactions.

## References

American Water Works Association. (2017). *AWWA leading practices in asset management report.* https://www.cityofhastings.org/assets/site/utl/documents/AWWA%20Case%20Study%20Intro%20170412.pdf

Anthony, S. D., Viguerie, S. P., Schwartz, E. I., & Van Landeghem, J. (2017). 2018 corporate longevity forecast: Creative destruction is accelerating. Innosight, Strategy and Innovation. https://www.innosight.com/insight/corporate-longevity-creative-destruction-is-accelerating/

Argyris, C. (2017). *Integrating the individual and the organization.* Taylor & Francis. https://doi.org/10.4324/9780203788417

Argyris, C., Putnam, R., & McLain Smith, D. (1985). *Action science.* Jossey-Bass Publishers.

Argyris, C., & Schön, D. A. (1974). *Theory in practice—Increasing professional effectiveness.* John Wiley & Sons.

Argyris, C., & Schön, D. A. (1996). *Organizational learning II theory, method, and practice.* Addison Wesley Publishing.

Binz, C., Harris-Lovett, S., Kiparsky, M., Sedlak, D. L., & Truffer, B. (2016). The thorny road to technology legitimation—Institutional work for potable water reuse in California. *Technological Forecasting and Social Change,* 103, 249–263.

BIO. (2015). *Optimising water reuse in the EU. Public consultation analysis report.* https://ec.europa.eu/environment/water/blueprint/pdf/BIO_Water%20Reuse%20Public%20Consultation%20Report_Final.pdf

BPMN. (2020). Business Process Modelling and Notation 2.0 standard. https://www.omg.org/spec/BPMN/2.0/PDF

Cameron, K. S., & Quinn, R. E. (1999). *Diagnosing and changing organizational culture.* Addison-Wesley.

Capra, F., & Luisi, P. L. (2019). *The systems view of life. A unifying version.* 10th printing. Cambridge University Press.

Colton, R. D. (2020). *The affordability of water and wastewater service in twelve U.S. cities: A social, business, and environmental concern.* The Guardian. https://www.theguardian.com/environment/2020/jun/23/full-report-read-in-depth-water-poverty-investigation

The Conversation. (2017). When a river is a person. https://theconversation.com/when-a-river-is-a-person-from-ecuador-to-new-zealand-nature-gets-its-day-in-court-79278

Danilenko, A., Dickson, E., & Jacobsen, M. (2010). *Climate change and urban water utilities: Challenges and opportunities.* Water Sector Board of the World Bank Sustainable Development Network. www.worldbank.org/water

Earley, C. P. (2003). *Cultural intelligence: Individual interactions across cultures.* Stanford University Press.

Eisenstein, C. (2015). *The more beautiful world our hearts know is possible.* North Atlantic Books.

Estés, C.P. (2008). *Women who run with the wolves.* Rider & Co.

eThekwini Municipality. (2020). https://www.gov.za/about-government/contact-directory/kzn-municipalities/kzn-municipalities/ethekwini-metropolitan

Frankl, V. (2006). *Man's search for meaning.* Beacon Press.

Fuller, R. B. (1969). *Operating manual for spaceship earth.* Lars Müller Publishers.

Gallup. (2017). *State of the global workplace.* Employee Engagement Insights for Business Leaders Worldwide. https://www.gallup.com/workplace/257552/state-global-workplace-2017.aspx

Goleman, D. (1996). *Emotional intelligence: Why it can matter more than IQ.* Bantam Books.

Goleman, D. (2006). *Social intelligence: Beyond IQ, beyond emotional intelligence.* Bantam Trade Paperback.

Grant, A. (2016). *Originals: How non-conformists move the world.* Penguin Books.

Greiner, L. E. (1998, May/June). Evolution and revolution as organizations grow. *Harvard Business Review.*

Haidt, J. (2012). *The righteous mind: Why good people are divided by politics and religion.* Pantheon Publishing.

Harris-Lovett, S. R., Binz, C., Sedlak, D. L., Kiparsky, M., & Truffer, B. (2015). Beyond user acceptance: A legitimacy framework for potable water reuse in California. *Environmental Science & Technology, 49*(13), 7552–7561.

Hayashi, A. (2021). *Social presencing theater. The art of making a true move.* PI Press.

Hildebrandt, S., & Stubberup, M. (2012). *Sustainable leadership: Leadership from the heart.* Copenhagen Press.

Howard, G., Calow, R., Macdonald, A., & Bartram, J. (2016). Climate change and water and sanitation: Likely impacts and emerging trends for action. *Annual Reviews Environmental Resources, 41*, 253–276. https://doi.org/10.1146/annurev-environ-110615-085856

Hübl, T., & Avritt, J. (2021). *Healing collective trauma. A process for integrating our intergenerational and cultural wounds.* Sounds True Inc.

Ingildsen, P. (2020). *Water stewardship.* IWA Publishing. https://iwaponline.com/ebooks/book-pdf/701561/wio9781789060331.pdf

Ingildsen, P., & Olsson, G. (2016). *Smart water utilities.* IWA Publishing. https://iwaponline.com/ebooks/book/11/Smart-Water-Utilities-Complexity-Made-Simple

Institute of Public Works Engineering Australia. (2020). *International infrastructure management manual.* https://www.ipwea.org/publications/ipweabookshop/iimm

IRENA. (2020). *Global renewables outlook.* International Renewable Energy Agency. https://www.irena.org/publications/2020/Apr/Global-Renewables-Outlook-2020

Jordi, A. (2015, October). Legitimacy—The key to successful implementation. *EAWAG Aquatic Research News.* https://www.eawag.ch/fileadmin/Domain1/News/User_Acceptance_englisch.pdf

Kahneman, D. (2011). *Thinking, fast and slow.* Farrar, Strauss, and Giroux.

Kahneman, D., & Tversky, A. (1979). Prospect theory: An analysis of decision under risk. *Econometrica, 47*(2), 263–291.

Kane, G. C., Phillips, A. N., Copulsky, J. R., & Andrus, G. R. (2019). *The technology fallacy, how people are the real key to digital transformation.* MIT Press.

Kane, J. W., & Tomer, A. (2019). *Shifting into an era of repair: U.S. infrastructure spending trends.* Brookings Institute. https://www.brookings.edu/research/shifting-into-an-era-of-repair-us-infrastructure-spending-trends/

Kant, I. (1996). An answer to the question: "What is enlightenment?" In M. J. Gregor (Ed. & Trans.), *Immanuel Kant: Practical philosophy.* Cambridge University Press. https://www.marxists.org/reference/subject/ethics/kant/enlightenment.htm (Original work published 1784)

Kotter, J. P. (1996). *Leading change.* Harvard Business School Publishing.

Laloux, F. (2014). *Reinventing organizations: A guide to creating organizations inspired by the next stage of human consciousness.* Nelson Parker. http://www.reinventingorganizations.com/uploads/2/1/9/8/21988088/140305_laloux_reinventing_organizations.pdf

Lencioni, P. (2002). *The five dysfunctions of a team.* Jossey-Bass.

Liner, B., de Monsabert, S., & Morley, K. (2012). Strengthening social metrics within the triple bottom line of sustainable water resources. *World Review of Science, Technology and Sustainable Development,* 9(1), 74.

Lovelock, J. (2016). *Gaia. A new look at life on Earth.* Oxford Landmark Science.

March, J. G. (1994). *A primer on decision making: How decisions happen.* Free Press.

Markova, D., & McArthur, A. (2015). *Collaborative intelligence—Thinking with people who think differently.* Penguin Random House.

Maslow, A. H. (1943). A theory of human motivation. *Psychological Review, 50*(4), 370–396. https://doi.org/10.1037/h0054346

McMillan, C., & Overall, J. (2016). Wicked problems: Turning strategic management upside down. *Journal of Business Strategy.*

Naess, A. (2008). *Ecology of wisdom.* Counterpoint.

NAS. (2012). *Water reuse: Potential for expanding the nation's water supply through reuse of municipal wastewater.* National Academies Press.

Olsson, G. (2015). *Water and energy—threats and opportunities* (2nd ed). IWA Publishing.

Olsson, G. (2018). *Clean water using solar and wind: Outside the power grid.* IWA Publishing. https://iwaponline.com/ebooks/book/738/Clean-Water-Using-Solar-and-Wind-Outside-the-Power

Olsson, G., & Ingildsen, P. (2020). Process control. In G. Chen, M. van Loosdrecht, G. Ekama, & D. Brdjanovic (Eds.), *Biological wastewater treatment—Principles, modelling and design* (2nd ed.). IWA Publishing.

Porges, S. W. (2011). *The polyvagal theory. Neurophysiological foundations of emotions, attachment, communication and self-regulation.* W. W. Norton & Company.

Raworth, K. (2017). *Doughnut economics. Seven ways to think like a 21st century economist.* Penguin, Random House.

REN21. (2020). *Renewables 2020: Global status report.* Ren21 Secretariat, UN Environment Programme. https://www.ren21.net/reports/global-status-report/

Rittel, H. W., & Webber, M. M. (1973). Dilemmas in a general theory of planning. *Policy Sciences, 4*(2), 155–169.

Robbins, S. P., & Judge, T. A. (2007). *Organizational behavior* (12th ed.). Pearson Prentice Hall.

Salovey, P., & Mayer, J. D. (1990). Emotional intelligence. *Imagination, Cognition, and Personality,* 9(3), 185–211.

Sanford, C. (2017). *The regenerative business: Redesign work, cultivate human potential, achieve extraordinary outcomes.* John Murray Press.

Schein, E., & Schein, P. (2017). *Organizational culture and leadership* (5th ed.). John Wiley & Sons.

Schumacher, E. F. (2011). *Small is beautiful: A study of economics as if people mattered.* Random House. (Originally published in 1973)

Sedlak, D. (2014). *Water 4.0. The past, present, and future of the world's most vital resource.* Yale University Press.

Smart. (n.d.-a.) In *Merriam-Webster's online dictionary.* https://www.merriam-webster.com/dictionary/smart

Smart. (n.d.-b.) In *Oxford learner's online dictionary.* https://www.oxfordlearnersdictionaries.com/us/definition/english/smart_1

Smart Water Company of the Year. (2018). https://globalwaterawards.com/2018-smart-water-company-of-the-year/

Smart Water Utilities. (2019). https://www.smart-water-utilities.com

Sønderlund, A. L., Smith, J. R., Hutton, C. J., Zoran, K., & Savic, D. (2016). Effectiveness of smart meter-based consumption feedback in curbing household water use: Knowns and unknowns. *Journal of Water Resources Planning and Management, 142*(12). https://ascelibrary.org/doi/pdf/10.1061/%28ASCE%29WR.1943-5452.0000703

Stam, J. J. (2017). *Fields of connection. The practice of business connections.* Systemic Books Publishing.

Stone, C. (2010). *Should trees have standing?* Oxford University Press. (Originally published in 1972)

SWAN. (2020). *What is a smart water network?* The Smart Water Networks Forum. https://www.swan-forum.com/what-is-a-smart-water-network/

Taylor, N. (1998). *Urban planning theory since 1945.* Sage Publications.

Thaler, R. H., & Sunstein, C. S. (2008). *Nudge.* Yale University Press.

Thames Water. (2020). Request a water meter. https://www.thameswater.co.uk/help/water-meters/getting-a-water-meterhttps (Accessed December 22, 2020)

UAIM. (2017). Utility Analysis and Improvement Methodology Workshop #2. Prepared for Water Environment & Reuse Foundation, LIFT for Management (LIFT17T16), Report WEFTEC, Chicago, IL.

UAIM. (2020, April). UAIM Phase 3—Year 1 Project Report, LIFT for Management, Water Research Foundation.

UN. (2020). *World water development report 2020—Water and climate change.* https://www.unwater.org/publications/world-water-development-report-2020/

UN Environment Programme. (2008). *Vital water graphics. An overview of the state of the world's fresh and marine waters. Increasing price with volume.* https://www.unep.org/resources/report/vital-water-graphics-overview-state-worlds-fresh-and-marine-waters

U.S. Chamber of Commerce. (2016). *From scarcity to abundance. Business solutions for a water constrained world.* U.S. Chamber of Commerce Foundation. https://www.uschamberfoundation.org/sites/default/files/Water%20Case%20Study%20Web%2012.15.pdf

U.S. Congressional Budget Office. (2018). *Public spending on transportation and water infrastructure, 1956 to 2017.* https://www.cbo.gov/publication/54539

Vernadsky, V. (1998). *The biosphere.* Copernicus Books. (Originally published 1926)

Vitasovic, Z. (2011). IT in wastewater utilities: Survey of end user view. *WEFTEC Proceedings, Session 21.* Water Environment Federation.

Vitasovic, Z., Haskins, S., & Barnett, M. W. (Principal investigators). (2021). *Utility analysis and improvement methodology.* The Water Research Foundation Project #4806. Water Research Foundation.

Vitasovic, Z., & Karimova, F. (2020). *Technology silos may be in your organization's cultural DNA.* AWWA/WEF Utility Management Conference, Anaheim, CA.

Vitasovic, Z., Olsson, G., Liner, B., Sweeney, M., & Abkian, V. (2015). Utility analysis and integration model. *Journal of American Water Works Association, 107*, 8.

Water Research Foundation. (2020). *LIFT for management—Utility analysis and improvement methodology Phase 3 Year 1 report* (WRF project 4806).

Water Research Foundation. (2022). *Definition, framework, and maturity assessment for intelligent water systems* (WRF project 5039).

Wiener, N. (1961). *Cybernetics.* MIT Press. (Originally published 1948)

# Index

## Endnotes

1. The WISE concept and approach are documented as a short report (https://www.accesswater.org/publications/-10048517/water-intrapreneurs-for-successful-enterprises--wise---a-vision-for-water-utilities) and a video (https://www.youtube.com/watch?v=aRErWwjqyPk).
2. Wiener (1961[1948]).
3. The UAIM Framework was originally proposed by Vitasovic et al. (2015).
4. Materials from the first 2017 Utility Management Conference are available in UAIM (2017), see https://9danalyticscom-my.sharepoint.com/:b:/g/personal/cello_9danalytics_com/EaMQ3WuRCepCnZFfH9CZLF8BBrQbzyay27QREjS4SXpzGw?e=8dNSbS.
5. Business process model (BPMN, 2020) describes the standard method and notation for business process modeling. See further Chapter 6.2. See https://www.omg.org/spec/BPMN/2.0/About-BPMN/.
6. The experiences after the third UAIM year are documented in UAIM (2020); see https://9danalyticscom-my.sharepoint.com/:b:/g/personal/cello_9danalytics_com/EQXvPVy-yep9AkbLfmFSjalUBQNDS2-VD6QSwXmbDjTmRZQ?e=ZMl3Nq.
7. The results of the UAIM project are documented in the final report Vitasovic et al. (2021).
8. See Schumacher (2011 [1973]).
9. See Ingildsen and Olsson (2016).
10. Twenty large utilities around the world (including three cities in the United States and Europe) have presented their perceptions, experiences, and approaches to addressing climate-related challenges of urban areas (Danilenko et al., 2010; Howard et al., 2016).
11. The International Water Association (IWA) has presented a white paper "Water for Smart Livable Cities," on how rethinking urban water management can transform cities of the future. Version 1.0, printed in October 2020, is downloadable at https://stateofgreen.com/en/publications/.
12. The term "smart" associated with using advanced technology is described in several other publications (Smart Water Company of the Year, 2018; Smart Water Utilities, 2019; SWAN, 2020). In several dictionaries, including Merriam-Webster (n.d.-a.), the word "smart" is equated with "intelligent." More recently, several types of "intelligence" have been identified including emotional intelligence (Salovey & Mayer, 1990; Goleman, 1996), social intelligence (Goleman, 2006), cultural intelligence (Earley, 2003), and collaborative intelligence (Markova & McArthur, 2015).
13. One aspect of smart systems is to integrate the complete urban water cycle. Detailed descriptions of smart systems are found in the book Ingildsen and Olsson (2016) and in Olsson and Ingildsen (2020).
14. The meaning of *wise*. Per Oxford dictionary (n.d.-b.), wise is defined as "able to make sensible decisions and give good advice because of the experience and knowledge that you have," and Merriam Webster (n.d.-a.) defines it as "characterized by *wisdom*: marked by deep understanding, keen discernment, and (a) capacity for sound judgment, (b) exercising or showing sound judgment." A smart or intelligent person or organization is not automatically wise. Wise is related to consideration, care, responsibility, and solicitude. Even the smartest individual or organization must learn or acquire wisdom. We seldom talk about

young people as wise. A smart young person or a brilliant scientist is not necessarily wise. However, gradually they may acquire wisdom. Consequently, a perfectly operated utility that satisfies effluent water requirement and customer needs is evidently smart but not necessarily wise.

15. The UN World Water Development Report (UN, 2020), titled *Water and Climate Change* aims at helping water communities to tackle climate change challenges.
16. Examples of how nature has been given legal rights are described for example in The Conversation (2017). Christopher Stone's game-changing work from 1972 on the legal rights of nature are described in his book (Stone, 2010).
17. Lencioni (2002) describes how trust is the most fundamental asset of any organization.
18. See the box "the feedback principle," Chapter 1.
19. The double-loop learning concept, an organization's ability to learn and change is formulated in Argyris et al. (1974, 1985, 1996, 2017).
20. Organizational culture is a system of shared assumptions, values, and beliefs that governs how people behave in organizations. Peter Drucker (1909–2005), one of the most influential thinkers on management, famously said that "*culture eats strategy for breakfast*." Edgar Schein (Schein & Schein, 2017), a leading researcher in this field, defines culture as "*a pattern of shared basic assumptions learned by a group as it solved its problems of external adaptation and internal integration.*" Sections 3.2, 5.3, and 5.4 include detailed discussions on this topic.
21. Details of the UAIM project, including the Water Sector Value Model (WSVM), are documented in Water Research Foundation (2020) materials.
22. See Capra and Luisi (2019).
23. The purpose of work—the "sense and purpose" may be illustrated by the old story of the bricklayers. The parable is rooted in an authentic story when the famous architect Christopher Wren was commissioned to rebuild St Paul's Cathedral in London after the great fire of 1666. Wren asked the first bricklayer what he was doing, and he replied "I'm a bricklayer. I'm working hard laying bricks to feed my family." Another bricklayer answered the same question: "I'm a cathedral builder. I'm building a great cathedral to The Almighty." The attitude to work is different if a person feels that the job is to complete spreadsheets for a program of flood risk management compared to the thinking that the job is saving a quarter of a million people from flooding. A similar sense of purpose can also be motivating the utility. Are we completing specific tasks for a water company, or are we part of bringing water to thousands of families?.
24. The Gallup survey is published in Gallup (2017).
25. The book Kotter (1996) is an important source of information to understand organizational change.
26. The survey of the 45 wastewater utilities is published by Vitasovic (2011).
27. The documentation of the research at Boston University and Deloitte concerning people and technology is published by Kane et al. (2019).
28. The relationship between the organization and technology in water sector utilities is described by Vitasovic & Karimova (2020).
29. The REN21 *Renewables Global Status Report* (REN21, 2020) together with IRENA *Global Renewables Outlook* (2020) provide the world's most comprehensive reports on renewables in 2020.

30. Issues related to water availability and treatment like desalination are described in (Olsson, 2018).
31. National Academy of Science (NAS, 2012) describes possibilities of expanded water reuse.
32. In the United States (population around 330 million), around 44,000 million $m^3$ of municipal wastewater is discharged every year. Approximately 17,000 million $m^3$ are discharged to an ocean or estuary. Reusing water would directly augment the nation's total water supply. In the European Union (population around 445 million), more than 40,000 million $m^3$ of wastewater is treated in the European Union every year, but only around 960 million $m^3$ of this treated wastewater is reused. There is high potential for increased water reuse, but awareness of the benefits of this technology is low. Stronger regulatory and financial incentives could help increasing reuse. BIO (2015) describes reuse of treated wastewater in Europe. U.S. Chamber of Commerce (2016) has documented water reuse in the United States. It is estimated that California may have some 30,000 million $m^3$ of potentially reusable water to be developed. To put that in perspective, the city of Los Angeles supplies approximately 630 million $m^3$ per year to its customers. The most recent water recycling surveys place the state at only 13% municipal reuse. The survey by U.S. Chamber of Commerce has shown substantial support for water reuse and acceptance of using recycled water for potable reuse purposes, regardless of drought conditions.
33. Sønderlund et al. (2016) and Thames Water (2020) provide experiences of smart water meters.
34. Building trust—a South African experience. The eThekwini Water and Sanitation, serving Durban, South Africa, can be a role model on how to build up trust between the utility and the users. In 1992, Durban had 1 million people living in the city proper and another 1.5 million people, almost all black, who had moved into shantytowns or were living in housing projects just outside the city. It was estimated that 42% of the region's water was wasted because of broken water pipes and mains, leaky toilets, and faulty plumbing. In two districts with a combined population of half a million, 87% of the water was being lost due to leaks and other wastage. A crash program was initiated. The utility repaired and replaced mains, installed water meters, replaced leaking toilets to low flush toilets, and retrofitted showerheads and taps. Tanks were installed in homes and apartments for the low-income people to provide 200 liters of water a day for free per household. As a result, six years later the water consumption was less, even as another 800,000 people had received service. The conservation measures paid for themselves within a year. The program created something very precious, trust in the utility. People got to know that the utility will act as soon as there is a leakage or a problem with water delivery. This has created a win-win situation between users and the utility.
35. *The Guardian* (Colton, 2020) discussed affordability of water.
36. Tariffs in Durban, South Africa. Naturally a low or zero tariff applied to the first liters consumed can enhance affordability. For example, Durban, South Africa, provides the first 6 $m^3$/month (200 liters/day) per household of water a day free of charge for property chargeable values of less than Rand250,000 (USD 14,500), a lifeline to many. For more expensive properties, the tariff is 21 Rand/$m^3$ (USD 1.2/$m^3$) from start. There is no fixed charge. For a consumption higher than 6 $m^3$/month, the tariff is the same for everybody. If the water use is higher than 30 $m^3$/month, any household must pay around 52 Rand/$m^3$ (USD 3.0/$m^3$). Obviously, the purpose is creating disincentives for overuse. So, the utility must

balance the right for everybody to get clean water and to collect revenues to cover costs. Subsidizing water strategies depends a lot on the relation between connected low-income and high-income households. Cross-subsidies from high-consumption (high-income) to low-consumption (low-income) households are effective only if enough customers use the higher blocks. eThekwini Municipality (2020) provide data from Durban, South Africa.

37. Figure 4.2 is published in Olsson (2015).
38. Data in Figure 4.3 are based on UNEP (2008).
39. The Maslow's Hierarchy of Needs was proposed by Abraham Maslow in his publication on human motivation, Maslow (1943).
40. The success of the Groundwater Replenishment System (GWRS) in Orange County depended convincingly on public trust in utility management and water experts, as documented by Harris-Lovett et al. (2015), Jordi (2015), and Binz et al. (2016).
41. "Water wisdom" and water stewardship is the key content of the book Ingildsen (2020).
42. The WISE concept and approach are documented as a short report (https://www.accesswater.org/publications/-10048517/water-intrapreneurs-for-successful-enterprises--wise---a-vision-for-water-utilities) and a video (https://www.youtube.com/watch?v=aRErWwjqyPk).
43. The book (Frankl, 2006) is a famous discussion about man's search for meaning.
44. How our decision making is governed is discussed, for example, by Kahneman and Tversky (1979), Thaler and Sunstein (2008), and Kahneman (2011).
45. The philosophy of Kant (1996 [1784]) is available online.
46. Jonathan Haidt (2012) describes the moral framework that governs our decisions. The critical importance of a person's moral framework—system of values is a fundamental basis in water stewardship (Ingildsen, 2020).
47. The rational planning model for decisions was defined by Taylor (1998), and the rational decision model was described by Robbins and Judge (2007). Multiple examples of rational decisions are found in control systems, for examples in water and wastewater systems (Ingildsen & Olsson, 2016). A descriptive view of decision science including human interactions is documented in March (1994).
48. Grant (2016) has documented the Polaroid.
49. The development of private company evolutions has been studied in detail by Greiner (1998).
50. Corporate longevity has been analyzed by Anthony et al. (2017). They conclude that the S&P 500 lifespans are shrinking.
51. The concept of "wicked problems" was proposed by Rittel and Webber (1973). McMillan and Overall (2016) further compared conventional and wicked problems as outlined in Table 5.5.
52. The two categories of key challenges for an organization are described in detail in Schein and Schein (2017).
53. See American Water Works Association (2017).
54. See U.S. Congressional Budget Office (2018).
55. Kane and Tomer (2019) analyzes the costs for operation and maintenance of U.S. public infrastructure.
56. Asset management is discussed in Institute of Public Works Engineering Australia (2020).
57. See Water Research Foundation (2020).
58. See https://sdgs.un.org/goals.

59. See https://a4ws.org.
60. See Raworth (2017).
61. See https://www.presencing.org.
62. See Hayashi (2021).
63. See Lovelock (2016).
64. See Schumacher (2011 [1973]).
65. Deep ecology is an environmental philosophy that promotes the inherent worth of all living beings regardless of their instrumental utility to human needs. See Naess (2008).
66. Clarissa Pinkola Estés is an American author and psychoanalyst. Her most famous book is *Women Who Run with the Wolves*, see Estés (2008).
67. Eisenstein (2015) asks: "In a time of social and ecological crisis, what can we as individuals do to make the world a better place?" He describes how the principle of interconnectedness helps to give a strong positive influence.
68. See Hildebrandt and Stubberup (2012).
69. Sanford (2017) emphasizes that everything in business begins with you as an individual taking full responsibility for your actions. See also the video https://www.youtube.com/watch?v=ZjgUs719d48.
70. See Hübl and Avritt (2021).
71. See Fuller (1969).
72. Vernadsky (1998 [1926]) used the term *The Biosphere* hypothesizing that life is the geological force that shapes the earth.
73. Polyvagal theory (*poly*- "many" + *vagal* "wandering") is a collection of evolutionary, neuroscientific, and psychological claims pertaining to the role of the vagus nerve in emotion regulation, social connection, and fear response. The vagus is the longest nerve of the autonomic nervous system in the human body and comprises sensory and motor fibers. See Porges (2011).
74. Jan Jacob Stam is an organizational consultant, see Stam (2017).
75. See Sedlak (2014).

IWA Publishing's authorised EU representative for General Product Safety Regulations is Diane D'Arras, 15 rue Duret, 75116 Paris, France, e-mail: safety@iwap.co.uk.

Printed and bound by CPI Group (UK) Ltd, Croydon, CR0 4YY
17/09/2026
02218815-0001